普通高等教育"十二五"规划教材

GIS 软件 SharpMap 源码详解及应用

陈 真 何 津 余 瑞 编著

北 京
冶金工业出版社
2012

内 容 提 要

本书对目前基于 C#语言开发的 GIS 开源项目——SharpMap 进行详细剖析、解释,以便 GIS 专业学生及 GIS 的初学者能学习和掌握 GIS 底层开发技术。全书分为 3 部分共计 11 章。第 1 部分讲解 SharpMap 源码,主要内容包括地图、地图控件、图层、绘制、样式、数据、几何对象等;第 2 部分介绍基于 SharpMap 的应用开发,主要内容包括两个 SharpMap 下载包中附带的 Windows 应用程序;第 3 部分介绍 SharpMap 系统扩展,主要内容包括数据源对象扩展及图层对象扩展。

本书可作为地理信息系统相关专业本科生"GIS 开发与设计"等课程的教学用书,也可供对 GIS 感兴趣的初学者及 GIS 工程技术人员阅读参考。

图书在版编目(CIP)数据

GIS 软件 SharpMap 源码详解及应用/陈真,何津,余瑞编著.
—北京:冶金工业出版社,2012.3
普通高等教育"十二五"规划教材
ISBN 978-7-5024-5869-0

Ⅰ.①G… Ⅱ.①陈… ②何… ③余… Ⅲ.①地理信息系统—应用软件,SharpMap GIS—程序分析—高等学校—教材
Ⅳ.①P208

中国版本图书馆 CIP 数据核字(2012)第 022779 号

出 版 人　曹胜利
地　　址　北京北河沿大街嵩祝院北巷 39 号,邮编 100009
电　　话　(010)64027926　电子信箱　yjcbs@cnmip.com.cn
责任编辑　廖　丹　美术编辑　李　新　版式设计　孙跃红
责任校对　石　静　责任印制　李玉山
ISBN 978-7-5024-5869-0
三河市双峰印刷装订有限公司印刷;冶金工业出版社出版发行;各地新华书店经销
2012 年 3 月第 1 版,2012 年 3 月第 1 次印刷
787mm×1092mm　1/16;19 印张;455 千字;291 页
39.00 元

冶金工业出版社投稿电话:(010)64027932　投稿信箱:tougao@cnmip.com.cn
冶金工业出版社发行部　电话:(010)64044283　传真:(010)64027893
冶金书店　地址:北京东四西大街 46 号(100010)　电话:(010)65289081(兼传真)
(本书如有印装质量问题,本社发行部负责退换)

前 言

在过去的十年间，GIS（Geographic Information System，地理信息系统）已经从一个高度专业化的领域发展为影响我们生活方方面面的技术，从道路导航到自然灾害管理，随处可见 GIS 的应用。直到几年前还只有少数研究人员、决策人员和政府职员在使用 GIS，现在几乎每个人都可以通过互联网，利用众多 GIS 开发商提供的各种免费工具创建个性化地图或对 GIS 数据进行叠置分析。随着智能手机的普及，移动 GIS 应用技术也将迅速融入人们的生活，为普通百姓日常生活提供便利。

近年来 GIS 发展势头迅猛，新技术层出不穷，软件频频升级，设备不断更新。要跟上 GIS 发展的潮流，就必须不断了解和掌握 GIS 最新的发展动态。这也对 GIS 开发人员提出了更高的要求，不仅要知其然（掌握基于商用 GIS 平台的二次开发技术），还要知其所以然（了解 GIS 底层开发技术），从而具备更高的综合素质及竞争能力。

当前 GIS 软件有商业 GIS 系统及开源 GIS 系统之分。GIS 商用软件功能强大，有完善的技术支持，提供封装好的、功能强大的类库，基于商用 GIS 库进行的二次开发效率高、难度低、资源丰富。但对于小型 GIS 开发人员，商用 GIS 价格过高，对于 GIS 学习者来说，由于商用 GIS 产品的源代码是保密的，无法深入学习 GIS 底层技术，开发技术含量低，难以深化对 GIS 的理解；而开源 GIS 是完全免费的，其源代码完全公开，可作为非商业用途使用，是开发小型 GIS 系统或 GIS 开发人员学习的宝贵资料。但开源 GIS 项目的帮助资料通常很少，且没有强大的技术支持。此外在众多 GIS 开源代码中，C 语言及 Java 平台的项目居多，.NET 平台的 GIS 开源项目较少。SharpMap 是基于 .NET 2.0 平台用 C#语言开发的 GIS 开源项目，是一套简单易用的小型 GIS 平台，扩展性强，可以用于开发网络或桌面 GIS 应用程序。它支持多种 GIS 数据格式，支持空间查询，可渲染出精美地图。本书针对 SharpMap 的核心模块进行详尽的讲解，目的是为广大 .NET 平台的 GIS 开发人员提供服务，尽可能解决 SharpMap 开发文档匮乏的问题。

由于编者水平和经验有限，书中欠妥之处在所难免，恳请专家和广大读者不吝赐教。最后向为本书提出宝贵建议的专家、学者和同仁表示感谢。

编　者

2011 年 12 月

目 录

第 1 章　概述 ·· 1

  1.1　SharpMap 的特性 ·· 2
  1.2　SharpMap 支持的 GIS 数据格式 ·· 2
  1.3　SharpMap 名称空间概述 ·· 3
  1.4　SharpMap 中用到的第三方库 ·· 3
  1.5　SharpMap 源代码下载 ··· 4
  复习思考题 ·· 4

第 2 章　地图 ·· 5

  2.1　Map 类概述 ··· 5
  2.2　SharpMap 执行过程 ··· 5
  2.3　SharpMap 创建地图示例 ·· 5
  2.4　Map 类 ··· 7
    2.4.1　Map 类的数据成员 ·· 7
    2.4.2　Map 类的属性 ·· 7
    2.4.3　Map 类的方法 ·· 14
    2.4.4　Map 类的事件 ·· 19
  复习思考题 ·· 19

第 3 章　地图控件 ·· 20

  3.1　Tools 枚举 ·· 20
  3.2　MapImage 类 ··· 20
    3.2.1　MapImage 类的数据成员 ·· 20
    3.2.2　MapImage 类的属性 ·· 21
    3.2.3　MapImage 类的方法 ·· 23
    3.2.4　MapImage 类的事件 ·· 31
  复习思考题 ·· 32

第 4 章　图层 ·· 33

  4.1　图层对象概述 ··· 33
  4.2　ILayer 接口 ·· 34
    4.2.1　ILayer 接口的属性 ··· 34

4.2.2 ILayer 接口的方法 ………………………………………………………… 34
4.3 Layer 抽象类 …………………………………………………………………… 34
  4.3.1 Layer 抽象类的属性 …………………………………………………… 34
  4.3.2 Layer 抽象类的方法 …………………………………………………… 36
  4.3.3 Layer 抽象类的事件 …………………………………………………… 37
4.4 ICanQueryLayer 接口 …………………………………………………………… 37
4.5 矢量图层 VectorLayer …………………………………………………………… 38
  4.5.1 VectorLayer 的属性 …………………………………………………… 38
  4.5.2 VectorLayer 的方法 …………………………………………………… 40
  4.5.3 VectorLayer 的事件 …………………………………………………… 45
4.6 注记图层 LabelLayer …………………………………………………………… 45
  4.6.1 LabelLayer 的属性 …………………………………………………… 45
  4.6.2 LabelLayer 的方法 …………………………………………………… 49
4.7 Layer 集合 ……………………………………………………………………… 56
  4.7.1 LayerCollection 的索引器 …………………………………………… 56
  4.7.2 LayerCollection 的方法 ……………………………………………… 57
复习思考题 …………………………………………………………………………… 58

第 5 章 绘制 …………………………………………………………………………… 59

5.1 ClipState 枚举 …………………………………………………………………… 59
5.2 LabelBox 类 ……………………………………………………………………… 60
5.3 Label 类 ………………………………………………………………………… 60
  5.3.1 Label 类的数据成员 ………………………………………………… 60
  5.3.2 Label 类的属性 ……………………………………………………… 61
  5.3.3 Label 类的方法 ……………………………………………………… 61
5.4 LabelCollisionDetection 类 …………………………………………………… 62
  5.4.1 LabelCollisionDetection 类的方法 ………………………………… 62
  5.4.2 LabelCollisionDetection 类的代理 ………………………………… 63
5.5 矢量数据渲染类 VectorRender …………………………………………………… 63
  5.5.1 VectorRender 类的数据成员 ………………………………………… 63
  5.5.2 VectorRender 类的属性 ……………………………………………… 64
  5.5.3 VectorRender 类的方法 ……………………………………………… 64
5.6 主题 ……………………………………………………………………………… 73
  5.6.1 ITheme 接口 ………………………………………………………… 73
  5.6.2 色彩混合类 ColorBlend ……………………………………………… 73
  5.6.3 GradientThemeBase 类 ……………………………………………… 79
  5.6.4 GradientTheme 类 …………………………………………………… 86
  5.6.5 CustomTheme 类 …………………………………………………… 87
复习思考题 …………………………………………………………………………… 88

第6章 样式 ... 89

6.1 矢量图层样式 VectorStyle ... 89
6.1.1 VectorStyle 的数据成员 ... 89
6.1.2 VectorStyle 的属性 ... 90
6.1.3 VectorStyle 的方法 ... 90
6.2 标注样式 LabelStyle ... 91
6.2.1 LabelStyle 的数据成员 ... 91
6.2.2 LabelStyle 的属性 ... 92
6.2.3 LabelStyle 的方法 ... 92
复习思考题 ... 93

第7章 数据 ... 94

7.1 空间数据库连接池技术 ... 94
7.1.1 数据连接对象 Connector ... 95
7.1.2 连接池管理 ConnectorPool ... 98
7.2 数据提供接口 IProvider ... 100
7.2.1 IProvider 的属性 ... 100
7.2.2 IProvider 的方法 ... 100
7.3 DbaseReader 类 ... 102
7.3.1 DbaseReader 类的数据成员 ... 102
7.3.2 DbaseReader 类的属性 ... 103
7.3.3 DbaseReader 类的方法 ... 104
7.4 数据提供者 ShapeFile ... 111
7.4.1 ShapeFile 的数据成员 ... 112
7.4.2 ShapeFile 的属性 ... 114
7.4.3 ShapeFile 的方法 ... 116
7.5 MsSql 类 ... 131
7.5.1 MsSql 类的数据成员 ... 131
7.5.2 MsSql 类的属性 ... 131
7.5.3 MsSql 类的方法 ... 133
7.6 其他 Provider 类 ... 144
7.7 FeatureDataSet 类 ... 144
7.7.1 FeatureDataSet 类的属性 ... 145
7.7.2 FeatureDataSet 类的方法 ... 145
7.8 FeatureDataTable 类 ... 148
7.8.1 FeatureDataTable 类的属性 ... 148
7.8.2 FeatureDataTable 类的方法 ... 148
7.8.3 FeatureDataTable 类的事件 ... 150

7.9　FeatureDataRow 类 ································· 150
　7.9.1　FeatureDataRow 类的属性 ················· 150
　7.9.2　FeatureDataRow 类的方法 ················· 151
复习思考题 ··· 151

第 8 章　几何对象 ······································ 152

8.1　几何对象抽象基类 Geometry ···················· 152
　8.1.1　Geometry 的属性 ···························· 152
　8.1.2　Geometry 的方法 ···························· 154
8.2　点对象 Point ·· 159
　8.2.1　Point 的数据成员 ····························· 159
　8.2.2　Point 的属性 ·································· 159
　8.2.3　Point 的方法 ·································· 161
8.3　复合点对象 MultiPoint ···························· 166
　8.3.1　MultiPoint 的属性 ···························· 166
　8.3.2　MultiPoint 的方法 ···························· 167
8.4　线状几何形状的抽象类 Curve ··················· 170
　8.4.1　Curve 的属性 ·································· 170
　8.4.2　Curve 的方法 ·································· 171
8.5　多边形 Polygon ····································· 171
　8.5.1　Polygon 的属性 ······························· 171
　8.5.2　Polygon 的方法 ······························· 173
8.6　外包矩形框 BoundingBox ························ 178
　8.6.1　BoundingBox 的属性 ························ 178
　8.6.2　BoundingBox 的方法 ························ 180
8.7　空间关系类 SpatialRelations ····················· 189
复习思考题 ··· 191

第 9 章　Windows 应用程序开发——WinFormSamples ··············· 192

9.1　数据 ··· 192
9.2　系统简介 ·· 193
9.3　代码分析 ·· 194
　9.3.1　主窗体代码 ···································· 194
　9.3.2　数据访问代码 ································· 197
复习思考题 ··· 210

第 10 章　Windows 应用程序开发——DemoWinForm ··············· 211

10.1　数据 ·· 211
10.2　系统简介 ··· 211

10.3 代码分析 ·· 212
  10.3.1 数据访问代码 ··· 212
  10.3.2 主窗体代码 ··· 212
复习思考题 ·· 219

第11章 数据源扩展与图层对象扩展 ·· 220

11.1 DataTablePoint 类 ·· 220
  11.1.1 DataTablePoint 类的数据成员 ··· 220
  11.1.2 DataTablePoint 类的属性 ·· 220
  11.1.3 DataTablePoint 类的方法 ·· 222

11.2 OgrProvider 类 ·· 227
  11.2.1 OgrProvider 类的数据成员 ··· 228
  11.2.2 OgrProvider 类的属性 ··· 229
  11.2.3 OgrProvider 类的方法 ··· 232

11.3 GdalRasterLayer 类 ·· 243
  11.3.1 GdalRasterLayer 类的数据成员 ·· 244
  11.3.2 GdalRasterLayer 类的属性 ··· 244
  11.3.3 GdalRasterLayer 类的方法 ··· 252

复习思考题 ·· 288

附录 书中多次引用的基本概念 ·· 289

参考文献 ·· 291

第1章 概 述

　　SharpMap 是一套简单易用的制图库，可以用于开发网络或桌面 GIS 应用程序。它支持多种 GIS 数据格式，支持空间查询，可渲染出精美的地图（见图 1 - 1）。SharpMap 是基

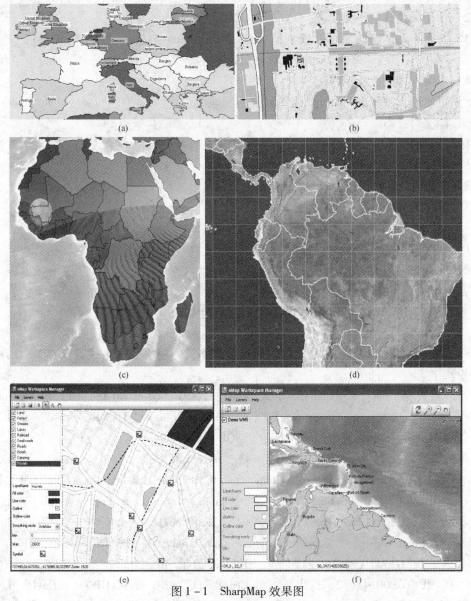

图 1 - 1　SharpMap 效果图
(a) 人口密度分级地图；(b) 城市街区地图；(c) 采用纹理及半透明效果的地图；
(d) ECW 栅格影像；(e) 采用地图控件的桌面应用系统；(f) WMS 服务数据

于微软的.NET 2.0 开发平台，用 C#语言开发的一套开源系统，受到 GNUL（GNU Lesser General Public License）许可保护。其网址为：http://sharpmap.iter.dk/ 和 http://www.codeplex.com/SharpMap。

1.1　SharpMap 的特性

SharpMap 具有以下特性：
（1）核心特性。
1）支持多种.NET 开发语言，包括 C#、VB.NET、C++/CLI 和 J#等；
2）使用属性数据作为注记；
3）符合 OpenGIS 的简单要素规范（OpenGIS Simple Features Specification），支持点、线、面、多点、多线、多面、几何对象集合等要素类型；
4）支持地图旋放及平移；
5）可使用代理（delegates）编制复杂主题地图；
6）能实时投影。
（2）网络地图。
1）支持使用 Http Handler 生成地图；
2）支持 AJAX 地图控件；
3）支持 WMS1.3.0 服务。
（3）扩展功能。
1）支持数据格式扩展；
2）支持图层扩展。

1.2　SharpMap 支持的 GIS 数据格式

SharpMap 支持的 GIS 数据格式包括矢量数据格式、栅格数据格式以及 WMS（网络地图服务）格式。

矢量数据格式有：ESRI Shape files format、PostGreSQL/PostGIS、OLEDB（points only）、Microsoft SQL Server、Oracle *、GPX *、MapInfo File *、TIGER *、S57 *、DGN *、CSV *、GML *、Interlis 1 *、Interlis 2 *、SQLite 和 ODBC *。

栅格数据格式有：Arc/Info ASCII Grid *、Arc/Info Binary Grid（.adf）*、Microsoft Windows Device Independent Bitmap（.bmp）*、ESRI.hdr Labelled *、ENVI.hdr Labelled Raster *、Envisat Image Product（.n1）*、Graphics Interchange Format（.gif）*、GMT Compatible netCDF *、GRASS Rasters *、TIFF/GeoTIFF（.tif）*、Erdas Imagine（.img）*、Idrisi Raster *、Image Display and Analysis（WinDisp）*、JPEG JFIF（.jpg）*、JPEG2000（.jp2,.j2k）*、JPEG2000（.jp2,.j2k）*、JPEG2000（.jp2,.j2k）*、JPEG2000（.jp2,.j2k）*、Erdas 7.x.LAN and.GIS *、Daylon Leveller Heightfield *、In Memory Raster *、NetCDF *、PCI.aux Labelled *、PCI Geomatics Database File *、Portable Network Graphics（.png）*、RadarSat2 XML（product.xml）*、

USGS SDTS DEM（CATD. DDF）∗、Raster Matrix Format（. rsw，. mtw）∗、SGI Image Format∗、USGS ASCII DEM（. dem）∗和 X11 Pixmap（. xpm）∗等。

WMS（网络地图服务）格式有：Version 1.0.0、Version 1.1.1、Version 1.3.0。网络地图服务（Web Map Service，WMS）是从地理信息动态产生，具有地理空间位置数据的地图。地图本身并不是数据，WMS 产生的地图一般以图像格式提供，如 PNG、GIF 或 JGPE；或按 SVG（Scalable Vector Graphics）或 Web CGM（Web Computer Graphics Metafile）格式提供基于矢量的图形元素。

1.3 SharpMap 名称空间概述

（1）SharpMap 名称空间只有一个类（Map 类），也是 SharpMap 的核心。

（2）Converts 名称空间提供数据转换服务。

（3）Forms 名称空间用于 Windows Form 编程，包含 MapImage 控件。MapImage 空间是一个简单的 User Control（用户控件），用于表达 Map 对象。

（4）Geometries 名称空间包括了 SharpMap 要使用到的各种几何类及其接口类，例如点、线、面等类，是 SharpMap 的基础之一。所有几何对象都继承自 Geometry 这个抽象类，其中定义了几何对象应该具备的公共操作，例如大小、ID、外接矩阵、几何运算等。

（5）Layers 名称空间包括了 ILayer 接口、Layer 集合类等，代表地图的图层。Layer 是一个抽象类，实现了 ILayer 接口。Layer 目前有 3 个子类，分别是 VectorLayer、LabelLayer 和 WmsLayer，分别代表 3 种不同数据类型的图层。

（6）Data 名称空间包含了与空间要素相关的类，如 FeatureDataSet、FeatureDataTable、FeatureDataRow、FeatureTableCollection 等。

（7）Data. Providers 名称空间包括了 IProvider 接口和 Shape 文件、PostGIS 数据访问实现代码。该名称空间为 SharpMap 提供数据读（写）支持，通过面向接口的设计，可以方便地扩展各类数据格式。

（8）Rendering 名称空间用于提供绘制空间数据的功能，目前包括矢量绘制对象和几个专题图绘制对象，该类将几何对象根据其 Style 设置，绘制到 System. Drawing. Graphics 对象上。

（9）Styles 名称空间主要提供了图层的样式类，例如线样式、点样式、填充样式等。

（10）Utilities 名称空间包括 Algorithms 类（目前仅实现了一个方法）、Providers 类（是数据提供对象 Provider 的一个 Helper 类，应用了反射机制）、Surrogates 类（主要用于系统的 Pen 和 Brush 的序列化）、Transform 类（提供了从图片坐标到地理坐标的互相变换，即桌面 GIS 的二次开发中经常使用的屏幕坐标和地理坐标的转换，主要用于地图的渲染、交互操作等）。

（11）Utilities. SpatialIndexing 名称空间用于对象的空间索引。

（12）Web 名称空间实现了 HttpHandler 和 Caching 类，用于网络环境。Web. UI. Ajax 提供对 Ajax 的支持。

1.4 SharpMap 中用到的第三方库

SharpMap 中用到的第三方库如下：

（1）ProjNET。这是一套基于微软. NET2.0 的开源地图投影库，支持基准面转换；支持大地坐标系、地心坐标系以及投影坐标系。其网址为：http：//projnet. codeplex. com/。

（2）BruTile。这是一套C#语言开发的支持GIS瓦片（tile）数据服务的开源代码库。其网址为：http://brutile.codeplex.com/。

（3）NetTopologySuite。这是一套快速、可靠的基于.NET的平台，用于处理GIS空间数据的拓扑关系，支持对二维线性几何形状进行拓扑运算。其网址为：http://code.google.com/p/nettopology。

（4）nunit.framework。这是一套基于.NET的单元测试平台。其网址为：www.nunit.org/。

1.5 SharpMap 源代码下载

打开网站 http://www.codeplex.com/SharpMap，在"Source Code"页面下，是项目各个版本的列表，显示各个版本的更新集编号（Change Set）、发布日期、下载次数以及更新说明。点击页面右边"Latest Version"文字下的链接 Download，可下载最新的项目打包文件。页面中点击更新集编号，可下载相应版本。下载后文件名为 sharpmap-×××××.zip，其中×××××是一个5位数字，表示当前下载文件的更新号。本书使用版本的更新集编号是78712，这是一个较稳定的版本，会与最新版有所不同，请读者注意。

下载的压缩文件中，有三个目录：一个是 Branches，为 SharpMap 的一些分支版本；一个叫 BuildProcessTemplates，是一些预定义的项目编译模板；第三个目录 Trunk，是包含项目代码的主目录，也是本书讲述的内容。

Trunk 目录下，有三个 Visual Studio 项目文件，即 SharpMap.sln、SharpMap.VS2008.sln、SharpMap.VS2010.sln，分别对应 Visual Studio 2005、Visual Studio 2008、Visual Studio 2010 三个版本。本书使用的是 SharpMap.VS2010.sln，建议使用 VS2010。

Trunk 目录还包含如下子目录：

（1）ExternalReferences。它是项目用到的第三方库文件。

（2）SharpMap。它是核心项目，包括 Map、Layer、Data、Geometries、Rendering、Style 等主要内容。

（3）SharpMap.UI。它包括地图控件。

（4）SharpMap.Extensions。它包括一些数据、图层扩展对象。

（5）UnitTests。它用于管理单元测试。

（6）SharpMap.SqlServerSpatialObjects。空间数据库，底层使用 MS SQL Server 的空间对象机制。

（7）PostgisDemoDb、MsSqlSpatialDemoDb。它们是使用 PostGIS、MsSqlSpatila 空间数据库的例子项目。

（8）DemoWebSite、DemoWinForm、WinFormSamples。它们是三个综合实例，第一个为 Web 应用，其他两个为桌面系统。

（9）ExampleCodeSnippets。它是 SharpMap 的代码段项目。

复习思考题

1-1 SharpMap 的功能和特性是什么，它支持的主要数据格式有哪些？

1-2 SharpMap 主要名称空间的功能是什么？

第2章 地 图

2.1 Map 类概述

Map 是 Sharp 的核心类之一，位于 SharpMap 名称空间下。Map 是装载地图图层 Layer 的容器，通过创建 Map 类的实例来生成地图，它由包含 Layer 对象的集合组成。用户可以通过创建 Map 对象的实例来得到地图，并通过 GetMap 方法来得到渲染（render）后的地图图形。

2.2 SharpMap 执行过程

在分析 SharpMap 之前，应该知道它的执行逻辑过程。典型的例子就是加载一个 Shape 文件的过程以及当地图的显示范围变化时，系统将要执行什么操作来更新地图视图。

Shape 文件只是数据源的一种，所以有一个访问数据源的接口来支撑这个数据访问层，接口对外界暴露一些访问数据的函数。毕竟在系统的高层（地图和图层层次上）并不会关注数据源来自哪里。这些我们会在第 7 章数据中做详细的分析。

2.3 SharpMap 创建地图示例

我们来看一下如何构造一个地图对象，完整代码如下：

```
SharpMap.Map myMap = new SharpMap.Map();
myMap.MinimumZoom = 100;
myMap.BackgroundColor = Color.White;
SharpMap.Layers.VectorLayer myLayer = new SharpMap.Layers.VectorLayer("My layer");
string shpLayerFullPath = @"C:\data\Lotlines.shp";
myLayer.DataSource = new ShapeFile(shpLayerFullPath);
myLayer.FillStyle = new SolidBrush(Color.FromArgb(240,240,240));
myLayer.OutlineStyle = new Pen(Color.Blue, 1);
myLayer.Style.Line.Width = 2;
myLayer.Style.Line.Color = Color.Black;
myLayer.Style.Line.EndCap = System.Drawing.Drawing2D.LineCap.Round;
myLayer.Style.Line.StartCap = layRailroad.LineStyle.EndCap;
myLayer.Style.Line.DashPattern = new float[] { 4.0f, 2.0f };
myLayer.Style.EnableOutline = true;
myLayer.SmoothingMode = System.Drawing.Drawing2D.SmoothingMode.AntiAlias;
```

```
myLayer.MaxVisible = 40000;
myMap.Layers.Add(myLayer);
myMap.Center = new SharpMap.Geometries.Point(725000,6180000);
myMap.Zoom = 1200;
myMap.Size = new System.Drawing.Size(300,200);
System.Drawing.Image imgMap = myMap.GetMap();
```

从上面的简单示例中我们可以看到，要构造一个完整的地图对象，SharpMap 需要经过以下几个步骤：

第一步，实例化一个 Map 对象，作为地图的容器。

```
SharpMap.Map myMap = new SharpMap.Map();
```

第二步，设置地图显示属性。

```
myMap.MinimumZoom = 100;//最小地图宽度为100
myMap.BackgroundColor = Color.White;//背景颜色设置为白色
```

第三步，实例化图层，将这个图层加入到地图对象中。

```
SharpMap.Layers.VectorLayer myLayer = new SharpMap.Layers.VectorLayer("My layer");
```

第四步，设置图层的数据源。

```
string shpLayerFullPath = @"C:\data\Lotlines.shp";
myLayer.DataSource = new ShapeFile(shpLayerFullPath);
```

第五步，设置图层的渲染样式。

将图层的填充样式设置成灰色 Color.FromArgb（240，240，240），用的是 .NET 的画刷类 SolidBrush。

```
myLayer.FillStyle = new SolidBrush(Color.FromArgb(240,240,240));
```

将外框线设置成蓝色，使用的是 .NET 的画笔类 Pen。

```
myLayer.OutlineStyle = new Pen(Color.Blue,1);
```

设置线条的宽度为两个像素。

```
myLayer.Style.Line.Width = 2;
```

设置线条的颜色是黑色。

```
myLayer.Style.Line.Color = Color.Black;
```

设置线的终止线头为圆头，使用 System.Drawing.Drawing2D 名称空间下的线头。

```
myLayer.Style.Line.EndCap = System.Drawing.Drawing2D.LineCap.Round;
```

设置线的起始线头为圆头。

```
myLayer.Style.Line.StartCap = layRailroad.LineStyle.EndCap;
```

设置虚线样式。

```
myLayer. Style. Line. DashPattern = new float[ ] { 4.0f, 2.0f };
```
允许外包框。
```
myLayer. Style. EnableOutline = true;
```
设置平滑模式为抗锯齿。
```
myLayer. SmoothingMode = System. Drawing. Drawing2D. SmoothingMode. AntiAlias;
```
最大显示比例尺度为40000。
```
myLayer. MaxVisible = 40000;
```
第六步，加载图层，将图层载入地图中。
```
myMap. Layers. Add( myLayer);
```
第七步，设置显示地图属性。
设置地图中心为点（725000, 6180000）。
```
myMap. Center = new SharpMap. Geometries. Point(725000, 6180000);
```
地图的宽度为1200。
```
myMap. Zoom = 1200; //Set zoom level
```
设置地图的屏幕大小。
```
myMap. Size = new System. Drawing. Size(300,200);
```
第八步，渲染地图，调用 Map 对象的 GetMap() 方法渲染地图。
```
System. Drawing. Image imgMap = myMap. GetMap();
```
以上代码中的用法将在本书后面的内容中进行详细的解释。

2.4 Map 类

2.4.1 Map 类的数据成员

NumberFormatEnUs 是 Map 类的数据成员，是 Map 类中定义的一个静态变量。它的值反映了当前 SharpMap 使用的数字格式信息。完整定义代码如下：

```
public static NumberFormatInfo NumberFormatEnUs = new CultureInfo ("en - US", false)
. NumberFormat;
```

它把数字格式设置为"美国"数字，并使用默认的格式（useUserOverride 设置为 false）。

2.4.2 Map 类的属性

2.4.2.1 Layers

Layers 获取或者设置地图图层 Layer 对象的集合，它是一个 LayerCollection 的集合，具

有增加、删除、定位图层的功能,代码如下:

```
private LayerCollection _Layers;
public LayerCollection Layers
{
    get { return _Layers; }
    set
    {
        int iBefore = 0;
        if ( _Layers ! = null )
            iBefore = _Layers. Count;
        _Layers = value;
        if ( value ! = null )
        {
            if ( LayersChanged ! = null )
                LayersChanged( );
            if ( MapViewOnChange ! = null )
                MapViewOnChange( );
        }
    }
}
```

从以上代码可以看到,get 访问器直接返回地图几何对象 LayerCollection;在 set 访问器中,若设置的值不为空,分别检查 LayersChanged、MapViewOnChange 事件是否为空,若不为空则分别触发图层变化事件 LayersChanged( ) 和地图视图变化事件 MapViewOnChange( ),如用户注册了相应事件,相应事件处理函数会被调用。

2.4.2.2 MapTransform

MapTransform 为变换矩阵,其利用矩阵运算对地图图像进行变形,代码如下:

```
private System. Drawing. Drawing2D. Matrix _MapTransform;
internal System. Drawing. Drawing2D. Matrix MapTransformInverted;
public Matrix MapTransform
{
    get { return _MapTransform; }
    set
    {
        _MapTransform = value;
        if ( _MapTransform. IsInvertible )
        {
            MapTransformInverted = _MapTransform. Clone( );
            MapTransformInverted. Invert( );
        }
        else
            MapTransformInverted. Reset( );
```

从以上代码可以看到,在 set 访问器中,除了给_MapTransform 赋值外,如果_MapTransform 可逆(_MapTransform.IsInvertible),就将_MapTransform 精确复制后,复制给 MapTransformInverted,然后进行逆矩阵运算 MapTransformInverted.Invert();如果_MapTransform 不可逆,则重置 MapTransformInverted 为单位矩阵。

2.4.2.3 Center

Center 指在地图坐标系中,中心的位置,代码如下:

```
private Point _Center;
public Point Center
{
    get { return _Center; }
    set
    {
        _Center = value;
        if (MapViewOnChange ! = null)
            MapViewOnChange();
    }
}
```

Point 就是使用的 SharpMap.Geometries 名称空间下的 Point 类,这是一个 X、Y 值都是 double 类型的二维空间点(三维空间点是 SharpMap.Geometries 名称空间下的 Point3D);set 访问器除了给_Center 进行赋值外,还要触发地图视图变化事件 MapViewOnChange()。

2.4.2.4 Envelop

Envelop 能获取基于当前比例尺、地图中心、尺寸下地图的外围矩形,它是一个定义在 SharpMap.Geometries 名称空间下的 BoundingBox 对象,反映了"左上"、"右下"的矩形关系,这十分有利于相交、包含、相邻等空间拓扑关系的判断。当需要得到一个几何要素的 Envelop 时,只需要调用这个要素的 GetBoundingBox() 方法就可获得,代码如下:

```
public BoundingBox Envelope
{
    get
    {
        return new BoundingBox(
            new Point(Center.X - Zoom * .5, Center.Y - MapHeight * .5),
            new Point(Center.X + Zoom * .5, Center.Y + MapHeight * .5));
    }
}
```

这个外包框是在需要时计算而来的,左上角 X 就是地图中心 X 减去地图宽度的一半,左上角 Y 就是地图中心 Y 坐标值减去地图高度的一半;右下角同理。

2.4.2.5 BackColor

BackColor 表示地图背景颜色。当背景颜色改变后，检查 MapViewOnChange 事件是否为空，若不为空则触发 MapViewOnChange 事件刷新地图视图，具体代码如下：

```
private Color _BackgroundColor;
public Color BackColor
{
    get { return _BackgroundColor; }
    set
    {
        _BackgroundColor = value;
        if (MapViewOnChange != null)
            MapViewOnChange();
    }
}
```

2.4.2.6 MinimumZoom

MinimumZoom 为空间坐标系下地图最小显示宽度值，这个值可以为0，但不能小于0，否则抛出异常，代码如下：

```
private double _MinimumZoom;
public double MinimumZoom
{
    get { return _MinimumZoom; }
    set
    {
        if (value < 0)
            throw (new ArgumentException("Minimum zoom must be 0 or more"));
        _MinimumZoom = value;
    }
}
```

2.4.2.7 MaximumZoom

MaximumZoom 为空间坐标系下地图的最大显示宽度值，不能小于0，否则抛出异常，代码如下：

```
private double _MaximumZoom;
public double MaximumZoom
{
    get { return _MaximumZoom; }
    set
    {
        if (value <= 0)
            throw (new ArgumentException("Maximum zoom must larger than 0"));
```

2.4 Map 类

```
        _MaximumZoom = value;
    }
}
```

2.4.2.8　Zoom

Zoom 表示空间坐标系下地图的显示宽度（该属性应为 MapWidth 较好，也可与地图高度 MapHeight 对应），Zoom 的 set 访问器直接返回这个值；set 访问器中确保值在用户定义的最大最小值之间，同时，若地图视图事件不为空，则要触发 MapViewOnChange 事件，刷新地图视图，代码如下：

```
private double _Zoom;
public double Zoom
{
    get { return _Zoom; }
    set
    {
        if (value < _MinimumZoom)
            _Zoom = _MinimumZoom;
        else if (value > _MaximumZoom)
            _Zoom = _MaximumZoom;
        else
            _Zoom = value;
        if (MapViewOnChange != null)
            MapViewOnChange();
    }
}
```

2.4.2.9　PixelSize

PixelSize 表示一个像素所代表的坐标单位，它实际上就是一个像素所代表的实际宽度，具体代码如下：

```
public double PixelSize
{
    get { return Zoom/Size.Width; }
}
```

2.4.2.10　PixelWidth

PixelWidth 表示返回一个像素宽度所代表的坐标单位宽度，它实际上直接返回上面的 PixelSize 值，代码如下：

```
public double PixelWidth
{
    get { return PixelSize; }
}
```

2.4.2.11 PixelAspectRatio

PixelAspectRatio 表示像素的长宽比，可以理解成：PixelAspectRatio = 每像素所代表的长度/每像素所代表的宽度（或高度）。同样，set 访问器中需要判断长宽比是否小于或等于零，若是则抛出异常，提示不合法的长宽比，代码如下：

```
public double PixelAspectRatio
{
    get { return _PixelAspectRatio; }
    set
    {
        if (_PixelAspectRatio <= 0)
            throw new ArgumentException("Invalid Pixel Aspect Ratio");
        _PixelAspectRatio = value;
    }
}
```

2.4.2.12 PixelHeight

PixelHeight 为返回一个像素的高度所代表的坐标单位高度，代码如下：

```
public double PixelHeight
{
    get { return PixelSize * _PixelAspectRatio; }
}
```

2.4.2.13 MapHeight

MapHeight 为地图高度——在空间坐标系下的显示高度。这个值是计算出来的，是一个只读属性。Zoom/Size.Width 即是 PixelWidth（每像素宽度代表的坐标单位宽度）乘以 PixelAspectRatio（每像素高度代表的坐标单位高度）再乘以 Size.Height，代码如下：

```
public double MapHeight
{
    get { return (Zoom * Size.Height)/Size.Width * PixelAspectRatio; }
}
```

2.4.2.14 Size

获取和设置地图显示尺寸，使用的是 .NET System.Drawing 名称空间下的 Size 类：private Size _Size，代码如下：

```
public Size Size
{
    get { return _Size; }
    set { _Size = value; }
}
```

2.4.2.15 Disclaimer

Disclaimer 能获取和设置地图说明文字，它是一个字符串类型。在它的 set 访问器中，可以看出 Disclaimer 只能被设置一次，代码如下：

```
private String _disclaimer;
public String Disclaimer
{
    get { return _disclaimer; }
    set
    {
        //only set disclaimer if not already done
        if (String.IsNullOrEmpty(_disclaimer))
        {
            _disclaimer = value;
            //Ensure that Font for disclaimer is set
            if (_disclaimerFont == null)
                _disclaimerFont = new Font(FontFamily.GenericSansSerif, 8f);
        }
    }
}
```

2.4.2.16　DisclaimerFont

获取和设置地图声明使用的字库，使用的是 .NET 的 System.Drawing 名称空间下的 Font 类，代码如下：

```
private Font _disclaimerFont;
public Font DisclaimerFont
{
    get { return _disclaimerFont; }
    set
    {
        if (value == null) return;
        _disclaimerFont = value;
    }
}
```

2.4.2.17　DisclaimerLocation

DisclaimerLocation 能获取和设置地图说明的位置，1 表示右上角，2 表示左上角，3 表示左下角，0 表示右下角，定义如下：

```
private Int32 _disclaimerLocation;
/// <summary>
/// Location for the disclaimer
/// ///  2|1
/// ///  -+-
/// ///  3|0
/// </summary>
public Int32 DisclaimerLocation
```

```
        }
            get { return _disclaimerLocation; }
            set { _disclaimerLocation = value%4; }
    }
```

2.4.3　Map 类的方法

2.4.3.1　public Map（Size size）

public Map（Size size）为通过指定的尺寸实例化 Map 对象，在这个构造函数中，Layers = new LayerCollection（）实例化图层集合对象；BackColor = Color.Transparent 将地图背景设为系统的透明色；_MaximumZoom = double.MaxValue 将最大地图显示宽度设为 Double 的最大值；_MinimumZoom = 0 设最小地图宽度是 0；_Center = new Point（0, 0）将地图中心设为左上角；_Zoom = 1 将宽度值设为 1；_PixelAspectRatio = 1.0 将长宽比设为 1，即正方形显示；_MapTransform = new Matrix（）和 MapTransformInverted = new Matrix（）分别实例化了地图变换矩阵和其逆矩阵，代码如下：

```
public Map(Size size)
{
    Size = size;
    Layers = new LayerCollection();
    BackColor = Color.Transparent;
    _MaximumZoom = double.MaxValue;
    _MinimumZoom = 0;
    _MapTransform = new Matrix();
    MapTransformInverted = new Matrix();
    _Center = new Point(0, 0);
    _Zoom = 1;
    _PixelAspectRatio = 1.0;
}
```

2.4.3.2　public Map（）

它实际上调用 public Map（Size size），实例化一个默认大小为（300, 150）的 Map 对象，代码如下：

```
public Map() : this(new Size(300, 150))
{
}
```

2.4.3.3　GetMap（）

GetMap（）方法会得到一个渲染后的图像，最终的结果是.NET 提供的 Image 类型的栅格图像。它调用 RenderMap 进行地图的渲染，代码如下：

```
public Image GetMap()
{
    Image img = new Bitmap(Size.Width, Size.Height);
```

2.4 Map 类

```
Graphics g = Graphics.FromImage(img);
RenderMap(g);
g.Dispose();
return img;
}
```

2.4.3.4 RenderMap（Graphics g）

该方法为绘制地图。这个方法的关键在于遍历每个图层，调用_Layers[i].Render(g, this)，使每个图层调用自身的 Render 方法完成渲染，代码如下：

```
public void RenderMap(Graphics g)
{ if (g == null)
    throw new ArgumentNullException("g", "Cannot render map with null graphics object!");
  if (Layers == null || Layers.Count == 0)
        throw new InvalidOperationException("No layers to render");
  g.Transform = MapTransform;
    g.Clear(BackColor);
    g.PageUnit = GraphicsUnit.Pixel;
    for (int i = 0; i < _Layers.Count; i++)
    {if (_Layers[i].Enabled && _Layers[i].MaxVisible >= Zoom && _Layers[i].MinVisible < Zoom) _Layers[i].Render(g, this);
    }
    RenderDisclaimer(g);
  if (MapRendered != null) MapRendered(g); //Fire render event
}
```

2.4.3.5 FindLayer（string layername）

FindLayer（string layername）的返回值是 IEnumerable < ILayer > 类型，是一个可遍历的 Layer 对象的集合。它实现了 C#的迭代器模式（C#通过 yield 关键字，大大简化了迭代器模式的实现），代码如下：

```
public IEnumerable < ILayer > FindLayer(string layername)
{
    foreach (ILayer l in Layers)
        if (l.LayerName.Contains(layername))
            yield return l;
}
```

2.4.3.6 GetLayerByName（string name）

该方法通过图层的名称得到图层，遍历 Layer 集合，一旦找到名称相同的图层就会立即返回，代码如下：

```
public ILayer GetLayerByName(string name)
{
    for (int i = 0; i < _Layers.Count; i++)
```

```
if(String.Equals(_Layers[i].LayerName,name,StringComparison.InvariantCultureIgnoreCase))
        return _Layers[i];
    return null;
}
```

2.4.3.7　GetExtents ( )

该方法得到 Map 对象中所有图层的外接矩形进行 Join 操作后的外接矩形 BoundingBox 对象，BoundingBox 中定义了所有的空间关系操作，包括相交、包含、相邻、结合等空间关系运算，详见本书第 8 章，代码如下：

```
public BoundingBox GetExtents()
{
    if (Layers == null || Layers.Count == 0)
        throw (new InvalidOperationException("No layers to zoom to"));
    BoundingBox bbox = null;
    for (int i = 0; i < Layers.Count; i++)
    {
        if (bbox == null)
            bbox = Layers[i].Envelope;
        else
            bbox = bbox.Join(Layers[i].Envelope);
    }
    return bbox;
}
```

2.4.3.8　ImageToWorld (PointF p, Map map)

该方法提供屏幕坐标和地图坐标的相互转换功能。由于这种转换经常使用，设计者将其放在 SharpMap.Utilities 名称空间下的 Transform 类中，以静态方法调用，简单方便，而且在一定程度上会提高效率，代码如下：

```
public static Point MapToWorld(PointF p, Map map)
{
    return Transform.MapToWorld(p, this);
}
```

Transform 类中 MapToWorld 函数实现如下：

```
public static Point MapToWorld(PointF p, Map map)
{
    BoundingBox env = map.Envelope;
    return new Point(env.Min.X + p.X * map.PixelWidth, env.Max.Y - p.Y * map.PixelHeight);
}
```

2.4.3.9　PointF WorldToImage (Point p)

参见上面 ImageToWorld 函数，代码如下：

2.4 Map 类

```
public PointF WorldToImage(Point p)
{
    return Transform.WorldtoMap(p, this);
}
```

Transform 类中 WorldtoMap 函数实现如下：

```
public static PointF WorldtoMap(Point p, Map map)
{
    PointF result = new System.Drawing.Point();
    double Height = (map.Zoom * map.Size.Height)/map.Size.Width;
    double left = map.Center.X - map.Zoom * 0.5;
    double top = map.Center.Y + Height * 0.5 * map.PixelAspectRatio;
    result.X = (float)((p.X - left)/map.PixelWidth);
    result.Y = (float)((top - p.Y)/map.PixelHeight);
    return result;
}
```

2.4.3.10 ZoomToBox (BoundingBox bbox)

该方法将地图缩放至指定大小的矩形中，实现 GIS 中常用的拉框放大、缩小等。先将地图显示宽度设为矩形的宽度，再根据显示高度情况决定是否调整显示宽度。最后，重置地图中心，并引发地图视图变化事件，代码如下：

```
public void ZoomToBox(BoundingBox bbox)
{
    _Zoom = bbox.Width;
    if (Envelope.Height < bbox.Height)
        _Zoom *= bbox.Height/Envelope.Height;
    Center = bbox.GetCentroid();
}
```

2.4.3.11 ZoomToExtents ( )

该方法为全图显示的功能，可以实现地图复位的功能。它实际上调用的是 ZoomToBox 方法，将地图缩放到它本身的大小（本身的大小通过 GetExtents ( ) 方法来获取），代码如下：

```
public void ZoomToExtents()
{
    ZoomToBox(GetExtents());
}
```

2.4.3.12 RenderDisclaimer (Graphics g)

这个方法主要根据不同的位置，绘制地图说明字符串。下面代码中，首先声明一个 StringFormat 对象，用于封装文本布局信息（如对齐、文字方向和 Tab 停靠位），然后通过 VectorRenderer.SizeOfString 计算字符串的绘制尺寸。VectorRenderer.SizeOfString 使用了委托推断 C#2.0 的新语法，当新增或移除一个目标方法到委托列表的时候，编译器可以推断委托类型到相应的实例，然后根据字符串的位置 DisclaimerLocation 值来将其绘制出，代码如下：

```
private void RenderDisclaimer(Graphics g)
{
    StringFormat sf;
    //Disclaimer
    if (! String.IsNullOrEmpty(_disclaimer))
    {
        SizeF size = VectorRenderer.SizeOfString(g, _disclaimer, _disclaimerFont);
        size.Width = (Single)Math.Ceiling(size.Width);
        size.Height = (Single)Math.Ceiling(size.Height);
        switch (DisclaimerLocation)
        {
            case 0: //Right - Bottom
                sf = new StringFormat();
                sf.Alignment = StringAlignment.Far;
                g.DrawString(Disclaimer, DisclaimerFont, Brushes.Black,
                    g.VisibleClipBounds.Width,
                    g.VisibleClipBounds.Height - size.Height - 2, sf);
                break;
            case 1: //Right - Top
                sf = new StringFormat();
                sf.Alignment = StringAlignment.Far;
                g.DrawString(Disclaimer, DisclaimerFont, Brushes.Black,
                    g.VisibleClipBounds.Width, 0f, sf);
                break;
            case 2: //Left - Top
                g.DrawString(Disclaimer, DisclaimerFont, Brushes.Black, 0f, 0f);
                break;
            case 3: //Left - Bottom
                g.DrawString(Disclaimer, DisclaimerFont, Brushes.Black, 0f,
                    g.VisibleClipBounds.Height - size.Height - 2);
                break;
        }
    }
}
```

2.4.3.13 Dispose()

这个方法负责资源的清理，它遍历每个图层，如果图层实现了 IDisposable 接口（Disposable 是 .NET 资源管理的重要方式，可查阅相关数据关于 Dispose 模式的讲解），就调用 Dispose() 方法释放资源，最后清空图层集合，代码如下：

```
public void Dispose( )
{
    foreach ( Layer layer in Layers )
        if ( layer is IDisposable )
            ( ( IDisposable ) layer ). Dispose( );
    Layers. Clear( );
}
```

2.4.4　Map 类的事件

2.4.4.1　LayersChanged

LayersChanged 是图层变化事件。当图层添加、删除时会触发这个事件，这个事件是专门为图层级别的变化所设计的。在 Map 类中，就是 Layers 属性的 set 访问器，当 set 访问器添加新的图层后就会触发这个事件，代码如下：

```
public delegate void LayersChangedEventHandler( );
public event LayersChangedEventHandler LayersChanged;
```

2.4.4.2　MapViewOnChange

MapViewOnChange 是地图视图变化事件。当地图视图发生变化时，会触发这个事件。MapViewOnChange 算是 Map 类中最常用的事件之一，它和地图的显示息息相关。可以看到，Map 类中，Layers 属性的 set 访问器添加新的图层、背景颜色 BackColor 变化、地图中心 Center 改变、显示宽度 Zoom 变化等都会触发这个事件，此事件的代码如下：

```
public delegate void MapViewChangedHandler( );
public event MapViewChangedHandler MapViewOnChange;
```

2.4.4.3　MapRendered

MapRendered 是地图渲染后的事件。这个事件主要是当地图渲染完毕后调用，它在 Map 类中的 RenderMap（Graphics g）方法中，所有的渲染都完成后，触发这个事件，代码如下：

```
public delegate void MapRenderedEventHandler( Graphics g );
public event MapRenderedEventHandler MapRendered;
```

复习思考题

2-1　SharpMap 中地图对象是什么，怎样新建一个地图对象？

2-2　在 SharpMap 中如何进行地图操作？试理解地图放大、缩小、漫游、全视图显示等功能，画出算法流程图。

第3章 地图控件

SharpMap 的地图控件位于 SharpMap.UI 项目下，分为两个名称空间，即 SharpMap.Forms 及 SharpMap.Web.UI.Ajax，分别用于桌面应用系统及 Web 应用的地图控件。SharpMap 的 Web 地图控件 AjaxMapControl 支持 Ajax 技术（一种异步网页访问技术）。由于篇幅关系，本书没有涉及 Web 内容，有兴趣的读者请自行研究。

SharpMap.Forms 名称空间中有两个文件：一个文件为 MapImage.cs，定义了 MapImage 类，它派生自 System.Windows.Forms.PictureBox，提供了一些基本功能，如地图的缩放、移动和数据查询；另一个文件为 MapBox.cs，定义的地图控件类派生自 System.Windows.Forms.Control，该文件中定义的地图控件比 MapImage.cs 文件中定义的地图控件功能更多。有趣的是这两个文件之间的关系，它们是通过两个编译变量（UseMapBox、UseMapBoxAsMapImage）决定的。项目编译时如果没有定义上述两个变量，则 MapImage.cs 文件里定义的 MapImage 类会被编译。当定义了上述任一变量时，MapBox.cs 文件会被编译。如果编译变量是 UseMapBox，则地图控件的类名字为 MapBox，否则为 MapImage。由于 SharpMap 的文档很少，推测这种做法可能是为了保持与旧系统的兼容性。

由于 SharpMap 提供的例子程序中使用的是 MapImage.cs 文件定义的 MapImage 类，所以本书也就该类进行讲述。

3.1 Tools 枚举

SharpMap.Forms.Tools 是枚举对象，代表了地图控件的常用工具，分别是 Pan（移动）、ZoomIn（放大）、ZoomOut（缩小）、Query（查询）、None（无），代码如下：

```
public enum Tools
{
    Pan, //移动
    ZoomIn,//放大
    ZoomOut,//缩小
    Query,//查询
    None//无
}
```

3.2 MapImage 类

3.2.1 MapImage 类的数据成员

MapImage 的数据成员如下：

（1）_isCtrlPressed。_isCtrlPressed 指示键盘的 Ctrl 键是否按下。

　　private bool _isCtrlPressed = false;

（2）mousedrag。mousedrag 指示鼠标按下左键不放时的坐标点。

　　private System.Drawing.Point mousedrag;

（3）mousedragging。mousedragging 指示是否在进行鼠标拖放操作。

　　private bool mousedragging = false;

（4）mousedragImg。mousedragImg 指示鼠标拖放时当前显示地图的拷贝副本。

　　private Image mousedragImg;

3.2.2　MapImage 类的属性

3.2.2.1　WheelZoomMagnitude

WheelZoomMagnitude 为鼠标滚动一下的缩放因子，和 FineZoomFactor 属性一起用于计算滚轮滚动时地图的缩放比例。使用 Description 特性给这个属性添加了描述性的文字；使用 DefaultValue 设置属性的默认值为 2；使用 Category 特性表示此属性的类别，使用属性控件（PropertyGrid）时同类属性会显示在一起，代码如下：

```
private double _wheelZoomMagnitude = 2;

[Description("The amount which a single movement of the mouse wheel zooms by.")]
[DefaultValue(2)]
[Category("Behavior")]
public double WheelZoomMagnitude
{
    get { return _wheelZoomMagnitude; }
    set { _wheelZoomMagnitude = value; }
}
```

3.2.2.2　FineZoomFactor

FineZoomFactor 用于调整 WheelZoomMagnitude 属性，当 Ctrl 键按下时，WheelZoomMagnitude 会除以该值。该值大于 1 时，减小缩放因子；小于 1 时，增大缩放因子。负数时则相反。同样添加了 Description、DefaultValue、Category 三个特性，代码如下：

```
private double _fineZoomFactor = 10;

[Description("The amount which the WheelZoomMagnitude is divided by " +
    "when the Control key is pressed. A number greater than 1 decreases " +
    "the zoom, and less than 1 increases it. A negative number reverses " +
    "it.")]
[DefaultValue(10)]
```

```csharp
[Category("Behavior")]
public double FineZoomFactor
{
    get { return _fineZoomFactor; }
    set { _fineZoomFactor = value; }
}
```

3.2.2.3 Map

Map 实现对_Map 的封装，当设置新的地图后，调用 Refresh（）方法刷新地图。此属性使用了 DesignerSerializationVisibility 特性，代码如下：

```csharp
private Map _Map;

[DesignerSerializationVisibility(DesignerSerializationVisibility.Hidden)]
public Map Map
{
    get { return _Map; }
    set
    {
        _Map = value;
        if (_Map != null)
            Refresh();
    }
}
```

3.2.2.4 QueryLayerIndex

QueryLayerIndex 获取和设置地图的索引号，实现对数据成员_queryLayerIndex 的封装，代码如下：

```csharp
private int _queryLayerIndex;
public int QueryLayerIndex
{
    get { return _queryLayerIndex; }
    set { _queryLayerIndex = value; }
}
```

3.2.2.5 ActiveTool

ActiveTool 获取和设置当前激活的地图操作工具，可以是移动、缩小、放大、查询和无其中一个，当设置新的值后，根据工具的不同，鼠标样式也会不同，当为"漫游"操作时，鼠标变为手柄（Cursors.Hand）形状，其他工具时为十字（Cursors.Cross）形状。同时，若工具发生了变化（value != _Activetool），ActiveToolChanged 事件不为空，就触发 ActiveToolChanged 事件，代码如下：

```csharp
private Tools _Activetool;
```

3.2 MapImage 类

```
public Tools ActiveTool
{
    get { return _Activetool; }
    set
    {
        bool fireevent = (value ! = _Activetool);
        _Activetool = value;
        if (value = = Tools.Pan)
            Cursor = Cursors.Hand;
        else
            Cursor = Cursors.Cross;
        if (fireevent)
            if (ActiveToolChanged ! = null)
                ActiveToolChanged(value);
    }
}
```

3.2.3 MapImage 类的方法

3.2.3.1 MapImage()

MapImage() 为 MapImage 的默认构造函数，初始化一个大小为 PictureBox.Size 的地图对象，设置当前激活的工具为 Tools.None；设置鼠标按下 MouseDown、鼠标移动 MouseMove、鼠标起来 MouseUp、滚轮滚动 MouseWheel 事件的处理函数，并且设置鼠标的形状为十字 Cursors.Cross，设置双倍缓冲为 true，代码如下：

```
public MapImage()
{
    _Map = new Map(base.Size);
    _Activetool = Tools.None;
    base.MouseMove + = new System.Windows.Forms.MouseEventHandler(MapImage_MouseMove);
    base.MouseUp + = new System.Windows.Forms.MouseEventHandler(MapImage_MouseUp);
    base.MouseDown + = new System.Windows.Forms.MouseEventHandler(MapImage_MouseDown);
    base.MouseWheel + = new System.Windows.Forms.MouseEventHandler(MapImage_Wheel);
    Cursor = Cursors.Cross;
    DoubleBuffered = true;
}
```

3.2.3.2 Refresh()

Refresh() 刷新地图，当地图容器不为空时，设置地图的大小为 Pciturebox.Size，同时做出判断，调用 Map 对象的 GetMap 方法渲染地图，同时需要调用基类的 Refresh() 方法以便完成渲染工作。最后如果用户注册有 MapRefreshed 事件，就激发事件，代码

如下：

```
public override void Refresh()
{
    if(_Map != null)
    {
        _Map.Size = Size;
        if(_Map.Layers == null || _Map.Layers.Count == 0)
            Image = null;
        else
            Image = _Map.GetMap();
        base.Refresh();
        if(MapRefreshed != null)
            MapRefreshed(this, null);
    }
}
```

3.2.3.3　OnKeyDown（KeyEventArgs e）

OnKeyDown（KeyEventArgs e）为激发键盘按下事件 KeyDown 的函数，这里仅仅获取一个值，该值指示是否曾按下 Ctrl 键，设置_isCtrlPressed 值，同时激发基类的 OnKeyDown 函数激发基类的 KeyDown 事件，代码如下：

```
protected override void OnKeyDown(KeyEventArgs e)
{
    _isCtrlPressed = e.Control;
    Debug.WriteLine(String.Format("Ctrl: {0}", _isCtrlPressed));

    base.OnKeyDown(e);
}
```

3.2.3.4　OnKeyUp（KeyEventArgs e）

OnKeyUp（KeyEventArgs e）为激发键盘抬起事件 KeyUp 的函数，这里仅仅获取一个值，该值指示是否曾按下 Ctrl 键，设置_isCtrlPressed 值，同时激发基类的 OnKeyUp 函数激发基类的 KeyUp 事件，代码如下：

```
protected override void OnKeyUp(KeyEventArgs e)
{
    _isCtrlPressed = e.Control;
    Debug.WriteLine(String.Format("Ctrl: {0}", _isCtrlPressed));

    base.OnKeyUp(e);
}
```

3.2.3.5　OnMouseHover（EventArgs e）

OnMouseHover（EventArgs e）为激发 MouseHover 事件，这里调用 Focus() 方法为控

件设置输入焦点，如果输入焦点请求成功，则为 true，否则为 false。同时激发基类的 OnMouseHover 函数激发基类的 MouseHover 事件，代码如下：

```
protected override void OnMouseHover(EventArgs e)
{
    if(! Focused)
    {
        bool isFocused = Focus();
        Debug.WriteLine(isFocused);
    }
    base.OnMouseHover(e);
}
```

3.2.3.6 MapImage_Wheel(object sender, MouseEventArgs e)

该方法为鼠标滚动事件 MouseWheel 处理函数，这里根据鼠标滚轮的滚动幅度和_wheelZoomMagnitude、_fineZoomFactor 的值来确定地图对象_Map 的显示宽度。同时如果用户注册了 MapZoomChanged 事件，则激发事件，最后调用 Refresh() 方法刷新地图，代码如下：

```
private void MapImage_Wheel(object sender, MouseEventArgs e)
{
    if(_Map ! = null)
    {
        double scale = ((double)e.Delta / 120.0);
        double scaleBase = 1 + (_wheelZoomMagnitude / (10 * ((double)(_isCtrlPressed ? _fineZoomFactor : 1))));
        _Map.Zoom *= Math.Pow(scaleBase, scale);
        if(MapZoomChanged ! = null)
            MapZoomChanged(_Map.Zoom);
        Refresh();
    }
}
```

3.2.3.7 MapImage_MouseDown(object sender, MouseEventArgs e)

该函数为鼠标按下事件 MouseDown 处理函数。先判断是否按下左键，将当前鼠标点的坐标保存在 mousedrag 中，并继续引发 MouseDown 事件，代码如下：

```
private void MapImage_MouseDown(object sender, MouseEventArgs e)
{
    if(_Map ! = null)
    {
        if(e.Button == MouseButtons.Left) //dragging
            mousedrag = e.Location;
```

```
        if ( MouseDown ! = null)
            MouseDown( _Map. ImageToWorld( new System. Drawing. Point( e. X, e. Y) ) , e) ;
    }
}
```

3.2.3.8 MapImage_MouseMove (object sender, MouseEventArgs e)

该函数为鼠标移动事件 MouseMove 处理函数。先将当前点的坐标转换为世界坐标，继续激发 MouseMove 事件。然后将当前地图图像保存在 mousedragImage 中，设置 mousedragging 为 true，以免再次保存。然后根据当前选择的地图工具对地图进行操作；当选中的是漫游工具 Tools.Pan 时，调用 Graphics.DrawImageUnscaled 方法在指定的位置使用图像的原始物理大小绘制指定的图像，并将其保存到 Image 中，以便下次绘制；若选择的工具是放大或者缩小时，先计算出缩放比例和显示的矩形大小，最后同样调用 Graphics.DrawImageUnscaled 方法绘制地图，代码如下：

```
private void MapImage_MouseMove( object sender, MouseEventArgs e)
{
    if ( _Map ! = null)
    {
        Point p = _Map. ImageToWorld( new System. Drawing. Point( e. X, e. Y) ) ;
        if ( MouseMove ! = null)
            MouseMove( p, e) ;
        if ( Image ! = null && e. Location ! = mousedrag && ! mousedragging && e. Button
            = = MouseButtons. Left)
        {
            mousedragImg = Image. Clone( ) as Image;
            mousedragging = true;
        }
        if ( mousedragging)
        {
            if ( MouseDrag ! = null)
                MouseDrag( p, e) ;
            if ( ActiveTool = = Tools. Pan)
            {
                Image img = new Bitmap( Size. Width, Size. Height) ;
                Graphics g = Graphics. FromImage( img) ;
                g. Clear( Color. Transparent) ;
                g. DrawImageUnscaled( mousedragImg, new
                    System. Drawing. Point( e. Location. X - mousedrag. X, e. Location. Y -
                    mousedrag. Y) ) ;
                g. Dispose( ) ;
                Image = img;
            }
```

3.2 MapImage 类

```
        else if (ActiveTool = = Tools.ZoomIn || ActiveTool = = Tools.ZoomOut)
        {
            Image img = new Bitmap(Size.Width, Size.Height);
            Graphics g = Graphics.FromImage(img);
            g.Clear(Color.Transparent);
            float scale = 0;
            if (e.Y - mousedrag.Y < 0) //Zoom out
                scale = (float)Math.Pow(1 / (float)(mousedrag.Y - e.Y), 0.5);
            else //Zoom in
                scale = 1 + (e.Y - mousedrag.Y) * 0.1f;
            RectangleF rect = new RectangleF(0, 0, Width, Height);
            if (_Map.Zoom / scale < _Map.MinimumZoom)
                scale = (float)Math.Round(_Map.Zoom / _Map.MinimumZoom, 4);
            rect.Width *= scale;
            rect.Height *= scale;
            rect.Offset(Width / 2 - rect.Width / 2, Height / 2 - rect.Height
            / 2);
            g.DrawImage(mousedragImg, rect);
            g.Dispose();
            Image = img;
            if (MapZooming != null)
                MapZooming(scale);
        }
    }
}
```

3.2.3.9 MapImage_MouseUp (object sender, MouseEventArgs e)

该函数为鼠标抬起事件 MouseUp 处理函数,代码如下:

```
private void MapImage_MouseUp(object sender, MouseEventArgs e)
{
    if (_Map != null)
    {
```

若 MouseUp 事件不为空,先将当前点的坐标转换为世界坐标,并将其作为参数,触发 MouseUp 事件,代码如下:

```
if (MouseUp != null)
    MouseUp(_Map.ImageToWorld(new System.Drawing.Point(e.X, e.Y)), e);
```

再判断鼠标是否是左键按下,若是,则根据当前激活的工具,执行不同的操作,代码如下:

```
if (e.Button = = MouseButtons.Left)
{
```

当前激活的工具为缩小工具时，若不是正在进行拖拽操作，则调用_Map.ImageToWorld方法，将当前的鼠标点的图像坐标，转换为世界坐标，并将其设置为地图的显示中心点；若正在进行拖拽操作，则根据当前点Y坐标值e.Y和拖拽点的Y坐标值mousedrag.Y的差值是否大于零，计算并重新设置地图的比例尺，若小于零，说明是缩小操作，若大于零则为放大操作，最后刷新地图，代码如下：

```
if ( ActiveTool == Tools.ZoomOut)
{
    double scale = 0.5;
    if ( ! mousedragging)
    {
        _Map.Center = _Map.ImageToWorld(newSystem.Drawing.Point(e.X, e.Y));
        if (MapCenterChanged != null)
            MapCenterChanged(_Map.Center);
    }
    else
    {
        if (e.Y - mousedrag.Y < 0) //Zoom out
            scale = (float)Math.Pow(1 / (float)(mousedrag.Y - e.Y), 0.5);
        else //Zoom in
            scale = 1 + (e.Y - mousedrag.Y) * 0.1;
    }
    _Map.Zoom *= 1 / scale;
    if (MapZoomChanged != null)
        MapZoomChanged(_Map.Zoom);
    Refresh();
}
```

若当前激活的工具为放大工具时，操作方式同缩小时相似，请参考以上介绍，代码如下：

```
else if ( ActiveTool == Tools.ZoomIn)
{
    double scale = 2;
    if ( ! mousedragging)
    {
        _Map.Center = _Map.ImageToWorld(newSystem.Drawing.Point(e.X, e.Y));
        if (MapCenterChanged != null)
            MapCenterChanged(_Map.Center);
    }
    else
```

3.2 MapImage 类

```
            if (e.Y - mousedrag.Y < 0) //Zoom out
                scale = (float)Math.Pow(1 / (float)(mousedrag.Y - e.Y),
                0.5);
            else //Zoom in
                scale = 1 + (e.Y - mousedrag.Y) * 0.1;
        }
        _Map.Zoom *= 1 / scale;
        if (MapZoomChanged != null)
            MapZoomChanged(_Map.Zoom);
        Refresh();
    }
```

当前激活工具为漫游工具时，若正在进行拖拽操作，则重新计算并设置地图的显示中心，若 MapCenterChanged 事件不为空，则触发事件；若不是正在进行拖拽操作，则设置当前鼠标点为地图显示中心，同样的，若 MapCenterChanged 事件不为空，则触发事件，最后刷新地图，代码如下：

```
    else if (ActiveTool == Tools.Pan)
    {
        if (mousedragging)
        {
            System.Drawing.Point pnt = new System.Drawing.Point(Width / 2
                + (mousedrag.X - e.Location.X), Height / 2 + (mousedrag.Y -
                e.Location.Y));
            _Map.Center = _Map.ImageToWorld(pnt);
            if (MapCenterChanged != null)
                MapCenterChanged(_Map.Center);
        }
        else
        {
            _Map.Center = _Map.ImageToWorld(new System.Drawing.Point(e.X,
                e.Y));
            if (MapCenterChanged != null)
                MapCenterChanged(_Map.Center);
        }
        Refresh();
    }
```

若当前激活工具为查询工具时，首先判断查询图层的索引是否合法（_Map.Layers.Count > _queryLayerIndex && _queryLayerIndex > -1），若合法再继续操作。接着判断图层是否实现了 ICanQueryLayer 查询接口，若没有实现则提示无可查询图层；若实现了 ICanQueryLayer 查询接口，则继续操作，代码如下：

```
            else if ( ActiveTool = = Tools. Query)
            {
                if ( _Map. Layers. Count > _queryLayerIndex && _queryLayerIndex > - 1 )
                {
                    if ( _Map. Layers[ _queryLayerIndex ] is ICanQueryLayer)
                    {
```

将要查询的图层转换为 ICanQueryLayer 类型,保存于变量 layer 中,再判断地图的地图变换矩阵是否为单位矩阵,若是,则将当前鼠标点的坐标进行坐标转换,并将变换后的结果保存在 Point 类型的变量 pt 中,若不是单位矩阵,直接将坐标点保存在变量 pt 中,代码如下:

```
                        ICanQueryLayer layer = _Map. Layers[ _queryLayerIndex ] as ICanQueryLayer;
                        System. Drawing. Point pt;
                        if ( ! Map. MapTransform. IsIdentity)
                        {
                            System. Drawing. Drawing2D. Matrix mat =
                            Map. MapTransform;
                            mat. Invert( );
                            System. Drawing. Point[ ] pts = new
                            System. Drawing. Point[ 1 ];
                            pts[ 0 ] = new System. Drawing. Point( e. X, e. Y);
                            mat. TransformPoints( pts);
                            pt = pts[ 0 ];
                        }
                        else
                        {
                            pt = new System. Drawing. Point( e. X, e. Y);
                        }
```

调用 _Map. ImageToWorld 方法将 pt 点的坐标转换为世界坐标,调用其 GetBoundingBox 方法获取点的外包框矩形,然后调用 Grow 方法,生成一个大小为自身 5 倍的外包框矩形,作为查询范围,最后调用图层的 ExecuteIntersectionQuery 方法执行相交查询,将查询的结果保存在 FeatureDataSet 类型的变量 ds 中,若 MapQueried 事件不为空,且查询的结果不为空,将结果作为参数触发 MapQueried 事件,若查询的结果为空,则新生成一个 FeatureDataTable 对象作为参数,触发 MapQueried 事件,代码如下:

```
                        BoundingBox bbox = _Map. ImageToWorld( pt). GetBoundingBox( ). Grow( _Map. PixelSize * 5);
                        FeatureDataSet ds = new FeatureDataSet( );
                        layer. ExecuteIntersectionQuery( bbox, ds);
                        if ( MapQueried ! = null)
                        {
                            if ( ds. Tables. Count > 0)
                                MapQueried( ds. Tables[ 0 ]);
```

```
            else
                MapQueried(new FeatureDataTable());
        }
    }
    else
    MessageBox.Show("No active layer to query");
        }
    }
```

最后，判断临时存储的 mousedragImg 是否为空，若不为空则调用其 Dispose 方法回收资源后，将其设置为 null，并设置 mousedragging 值为 false，代码如下：

```
            if (mousedragImg ! = null)
            {
                mousedragImg.Dispose();
                mousedragImg = null;
            }
            mousedragging = false;
        }
}
```

3.2.4 MapImage 类的事件

3.2.4.1 MouseMove
MouseMove 为鼠标移动事件，当鼠标在地图上移动式触发，代码如下：

```
public delegate void MouseEventHandler(Point WorldPos, MouseEventArgs ImagePos);
public new event MouseEventHandler MouseMove;
```

3.2.4.2 MouseDown
MouseDown 为鼠标按下事件，当鼠标在地图上按下时触发，代码如下：

```
public new event MouseEventHandler MouseDown;
```

3.2.4.3 MouseUp
MouseUp 为鼠标（释放）起来事件，当鼠标（释放）起来时触发，代码如下：

```
public new event MouseEventHandler MouseUp;
```

3.2.4.4 MouseDrag
MouseDrag 为鼠标拖放事件，当鼠标拖放时触发，代码如下：

```
public event MouseEventHandler MouseDrag;
```

3.2.4.5 MapRefreshed
MapRefreshed 为地图刷新事件，当地图刷新后触发，代码如下：

```
public event EventHandler MapRefreshed;
```

3.2.4.6　MapZoomChanged

MapZoomChanged 为地图比例尺变化事件，当比例尺变化后触发，代码如下：

```
public event MapZoomHandler MapZoomChanged;
```

3.2.4.7　MapZooming

MapZooming 为地图缩放事件，当地图进行缩放后触发，代码如下：

```
public delegate void MapZoomHandler(double zoom);
public event MapZoomHandler MapZooming;
```

3.2.4.8　MapQueried

MapQueried 为地图查询事件，当完成地图查询后触发，代码如下：

```
public delegate void MapQueryHandler(FeatureDataTable data);
public event MapQueryHandler MapQueried;
```

3.2.4.9　MapCenterChanged

MapCenterChanged 为地图中心变化事件，当地图的中心变化后触发，代码如下：

```
public delegate void MapCenterChangedHandler(Point center);
public event MapCenterChangedHandler MapCenterChanged;
```

3.2.4.10　ActiveToolChanged

ActiveToolChanged 为地图工具变化事件，当所选的地图工具变化后触发，代码如下：

```
public delegate void ActiveToolChangedHandler(Tools tool);
public event ActiveToolChangedHandler ActiveToolChanged;
```

复习思考题

3-1　利用 SharpMap 提供的 MapImage 控件，编程实现地图的放大、缩小、漫游、全视图显示功能，体会利用地图控件编程的优点。

第 4 章 图 层

4.1 图层对象概述

Map 对象的一个重要组成部分就是图层（Layer），SharpMap 中所有图层都是从 ILayer 接口派生的，采用了面向接口的编程思想。虽然各种不同的图层，如矢量图层和标注图层相差很大，但它们都有共同的特点：图层的名称、最大最小可视范围、是否渲染、外包框等，而且图层都使用 Render 方法来渲染自己。为了共享代码，SharpMap 使用了 Layer 抽象类，它继承自 ILayer 接口，是其他类型图层类的基类。图 4-1 所示是 Layer 名称空间的对象图。

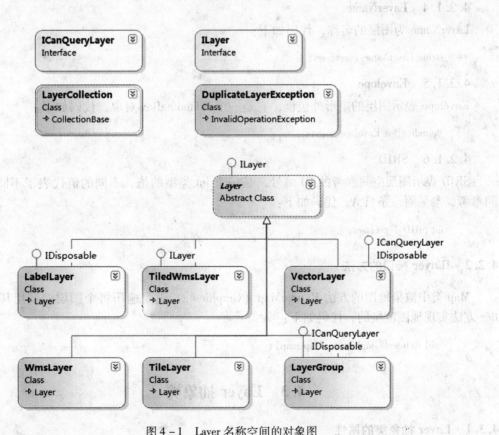

图 4-1　Layer 名称空间的对象图

由于篇幅关系，本书只对 ILayer、ICanQueryLayer、Layer、VectorLayer、LabelLayer、LayerCollection 等进行讲述，其他类原理相似，读者可自行研究。

4.2 ILayer 接口

4.2.1 ILayer 接口的属性

4.2.1.1 MinVisible

MinVisible 表示图层的最小可视比例尺，超过此值则不能显示，代码如下：

  double MinVisible { get; set; }

4.2.1.2 MaxVisible

MaxVisible 表示图层的最大可视比例尺，超过此值则不能显示，代码如下：

  double MaxVisible { get; set; }

4.2.1.3 Enabled

Enabled 决定图层是否可被渲染，是一个布尔值，代码如下：

  bool Enabled { get; set; }

4.2.1.4 LayerName

LayerName 为图层的名称，代码如下：

  string LayerName { get; set; }

4.2.1.5 Envelope

Envelope 表示图层的矩形外包框，它是一个 BoundingBox 对象，代码如下：

  BoundingBox Envelope { get; }

4.2.1.6 SRID

SRID 表示图层空间参考的 ID 编号，是一个 int 类型的值，不同的值代表了不同的空间参考，参见附录条目 A，代码如下：

  int SRID { get; set; }

4.2.2 ILayer 接口的方法

Map 类中渲染地图的方法 RenderMap（Graphics g）就是遍历每个图层，调用其 Render 方法实现地图渲染的，代码如下：

  void Render(Graphics g, Map map);

4.3 Layer 抽象类

4.3.1 Layer 抽象类的属性

4.3.1.1 Style

Style 表示图层的样式，它是一个 Style 类型值，详细信息可以参考第 6 章，相关代码

如下：

```
private Style _style;

public virtual Style Style
{
    get { return _style; }
    set { _style = value; }
}
```

4.3.1.2 MaxVisible

MaxVisible 表示最大可视范围，与 ILayer 接口的定义一致。这里 MaxVisible 使用 Style 对象的 MaxVisible 属性值作为该图层的 MaxVisible 值，代码如下：

```
public double MaxVisible
{
    get { return _style.MaxVisible; }
    set { _style.MaxVisible = value; }
}
```

4.3.1.3 MinVisible

MinVisible 表示最小可视范围，同样与 ILayer 接口的定义一致。这里 MinVisible 使用 Style 对象的 MinVisible 属性值作为图层的 MinVisible 值，代码如下：

```
public double MinVisible
{
    get { return _style.MinVisible; }
    set { _style.MinVisible = value; }
}
```

4.3.1.4 Enabled

Enabled 是一个布尔类型的值，指示该图层能否被渲染。同样，这里也是使用 Style 对象的 Enbled 属性值作为图层的 Enbled 值，代码如下：

```
public bool Enabled
{
    get { return _style.Enabled; }
    set { _style.Enabled = value; }
}
```

4.3.1.5 LayerName

LayerName 表示获取和设置图层的名称，它是一个 String 类型的值，代码如下：

```
private string _layerName;
public string LayerName
{
```

```
get { return _layerName; }
set { _layerName = value; }
}
```

4.3.1.6 Envelope

Envelope 表示图层的外包矩形，它是一个只读属性，是抽象函数，必须由子类实现，相关代码如下：

```
public abstract BoundingBox Envelope { get; }
```

4.3.1.7 CoordinateTransformation

CoordinateTransformation 表示设置或者获取图层的坐标变换，使用的是 Proj4.NET 开源类库的 ICoordinateTransformation 对象（参见附录条目 F），代码如下：

```
private ICoordinateTransformation _coordinateTransform;
public virtual ICoordinateTransformation CoordinateTransformation
{
    get { return _coordinateTransform; }
    set { _coordinateTransform = value; }
}
```

4.3.1.8 SRID

SRID 表示空间参考的 ID 编号，这是一个 int 类型的值，不同的空间编号代表着不同的空间参考，详细内容请参考附录条目 A，代码如下：

```
private int _srid = -1;
public virtual int SRID
{
    get { return _srid; }
    set { _srid = value; }
}
```

4.3.2 Layer 抽象类的方法

4.3.2.1 构造函数一

该函数实例化一个样式对象，以进行与样式相关的操作，代码如下：

```
protected Layer()
{
    _style = new Style();
}
```

4.3.2.2 构造函数二

该函数用指定的样式对象实例化图层的样式，传入的参数类型是 Style 类型。当然，不可以实例化 Layer 对象，因为 Layer 类是一个抽象类，代码如下：

```
public Layer(Style style)
{
    _style = style;
}
```

4.3.2.3 Render

Render（Graphics g，Map map），这个方法定义了图层渲染。可以看到，它调用的是 LayerRendered 所注册的事件处理函数，参数 g 是渲染所使用的 .NET Graphics 类型的对象，参数 map 表示所要渲染的地图，代码如下：

```
public virtual void Render(Graphics g, Map map)
{
    if(LayerRendered != null) LayerRendered(this, g); //Fire event
}
```

4.3.2.4 ToString()

此方法是重写了基类 Object 的 ToString() 方法，这里返回 LayerName 属性值，代码如下：

```
public override string ToString()
{
    return LayerName;
}
```

4.3.3 Layer 抽象类的事件

当地图渲染后触发 LayerRendered 事件，以供用户提供一些自定义操作，代码如下：

```
public delegate void LayerRenderedEventHandler(Layer layer, Graphics g);
public event LayerRenderedEventHandler LayerRendered;
```

4.4 ICanQueryLayer 接口

ICanQueryLayer 接口定义了图层查询操作，支持查询操作的图层（如 VectorLayer）从此接口派生。

（1）ExecuteIntersectionQuery。ExecuteIntersectionQuery 执行相交查询，获取指定的外包框对象 BoundingBox 内的要素数据集 FeatureDataSet，代码如下：

```
void ExecuteIntersectionQuery(BoundingBox box, FeatureDataSet ds);
```

（2）IsQueryEnabled。IsQueryEnabled 指示图层能否在 WmsServer 中查询。这个属性目前只能应用于 WMS，其他的情形只需调用 ExecuteIntersectionQuery() 方法即可完成查询。在以后的版本中，这个属性可能会被迁移到 ShaprpMap 的 SharpMap.Web.Wms.WmsServer 类中，代码如下：

```
bool IsQueryEnabled { get; set; }
```

4.5 矢量图层 VectorLayer

矢量图形与分辨率无关,也就是说,可以将它们任意缩放,按任意分辨率打印,而不丢失细节,也不会降低精度。

SharpMap 的矢量图层 VectorLayer 类实现了 ICanQueryLayer 接口和 IDisposable 接口,并继承自 Layer 抽象类。

4.5.1 VectorLayer 的属性

4.5.1.1 Style

Style 表示获取或者设置矢量图层的样式。这里使用的是 VectorStyle 对象,负责设置管理矢量图层的样式,详细信息可以参考第 6 章,代码如下:

```
public new VectorStyle Style
{
    get { return base.Style as VectorStyle; }
    set { base.Style = value; }
}
```

4.5.1.2 Envelope

Envelope 表示图层的外包矩形框,它是根据图层中各个要素的外包框来确定的,是一个只读属性。在 Get 访问器方法中,如果图层没有数据源(DataSource == null),就抛出异常;如果有数据源就判断数据源的状态,通过调用数据源的 GetExtents() 方法获得外包矩形框(关于数据源请参见第 7 章),同时,若坐标变换矩阵不为空,则对外包框进行坐标变换返回变换后的外包矩形框,代码如下:

```
public override BoundingBox Envelope
{
    get
    {
        if (DataSource == null)
            throw (new ApplicationException("DataSource property not set on layer '" + LayerName
                + "'"));

        bool wasOpen = DataSource.IsOpen;
        if (! wasOpen)
            DataSource.Open();
        BoundingBox box = DataSource.GetExtents();
        if (! wasOpen) //Restore state
            DataSource.Close();
        if (CoordinateTransformation ! = null)
```

4.5 矢量图层 VectorLayer

```
            return GeometryTransform.TransformBox( box, CoordinateTransformation. MathTrans-
form);
            return box;
        }
    }
```

4.5.1.3 SRID

SRID 表示获取或设置图层的空间参考的 ID 编号，这个编号是在数据源中定义的。这里直接使用数据源的编号作为图层的空间参考编号，代码如下：

```
public override int SRID
{
    get
    {
        if ( DataSource == null)
            throw (new ApplicationException("DataSource property not set on layer '" + LayerName
            + "'"));

        return DataSource.SRID;
    }
    set { DataSource.SRID = value; }
}
```

4.5.1.4 Theme

Theme 表示设置和获取图层的主题符号，若值为空，就会忽略图层的主题设置，详细信息可以参考第 5 章及第 6 章的相关章节，代码如下：

```
private ITheme _theme;
public ITheme Theme
{
    get { return _theme; }
    set { _theme = value; }
}
```

4.5.1.5 SmoothingMode

这个属性主要是设置图层的表达质量，它决定了直线、曲线、面状要素边缘的"平滑"效果，代码如下：

```
private SmoothingMode _smoothingMode;
public SmoothingMode SmoothingMode
{
    get { return _smoothingMode; }
    set { _smoothingMode = value; }
}
```

4.5.1.6 IsQueryEnabled

这个属性纯粹是为了和接口的设计相一致，其实在这里没有任何作用，代码如下：

```csharp
private bool _isQueryEnabled = true;
public bool IsQueryEnabled
{
    get { return _isQueryEnabled; }
    set { _isQueryEnabled = value; }
}
```

4.5.1.7 ClippingEnabled

ClippingEnabled 表示是否允许裁剪，若设置为 true，多边形要素就会在渲染之前被裁剪，因此，在只需要渲染要素的一部分时，将会大大提高渲染的速度，代码如下：

```csharp
private bool _clippingEnabled;
public bool ClippingEnabled
{
    get { return _clippingEnabled; }
    set { _clippingEnabled = value; }
}
```

4.5.1.8 DataSource

最后一个也是最重要的一个属性就是 DataSource 属性，它是一个 IProvider 对象，关于数据源的详细信息请参看第 7 章，这里仅仅设置或者返回数据源对象，代码如下：

```csharp
private IProvider _dataSource;
public IProvider DataSource
{
    get { return _dataSource; }
    set { _dataSource = value; }
}
```

4.5.2 VectorLayer 的方法

4.5.2.1 构造函数一 VectorLayer（string layername）

这个构造函数实例化一个图层，可以看到，构造函数中仅仅给出图层的名称和平滑效果，所以这样实例化的图层并没有实际的数据，在使用之前必须设置 DataSource 属性，以确定图层的数据源，代码如下：

```csharp
public VectorLayer(string layername)
    :base(new VectorStyle())
{
    LayerName = layername;
    SmoothingMode = SmoothingMode.AntiAlias;
}
```

4.5.2.2 构造函数二 VectorLayer (string layername, IProvider dataSource)

这个构造函数直接使用给定的名称和数据源来实例化一个图层对象。在内部，它实际上先调用构造函数一来生成一个空图层，然后设置图层的数据源，代码如下：

```
public VectorLayer(string layername, IProvider dataSource) : this(layername)
{
    _dataSource = dataSource;
}
```

4.5.2.3 RenderGeometry (Graphics g, Map map, Geometry feature, VectorStyle style)

此方法为渲染几何要素的方法，用给定的样式来渲染指定的几何要素。首先，得到指定几何要素 Geometry 的类型，根据不同的几何类型、相应的绘制格式（输入参数 style），调用 SharpMap.Rendering.VectorRenderer 类中对应的静态方法进行渲染（Rendering 类参见第 5 章），代码如下：

```
private void RenderGeometry(Graphics g, Map map, Geometry feature, VectorStyle style)
{
    GeometryType2 geometryType = feature.GeometryType;
    switch (geometryType)
    {
        case GeometryType2.Polygon:
            调用 VectorRenderer.DrawPolygon(…)方法
        case GeometryType2.MultiPolygon:
            调用 VectorRenderer.DrawMultiPolygon(…)方法

        …… 以下代码从略
    }
}
```

4.5.2.4 ExecuteIntersectionQuery (BoundingBox box, FeatureDataSet ds)

这是实现可查询图层 ICanQueryLayer 接口的方法，作用是执行相交查询，它直接调用数据源的 ExecuteIntersectionQuery () 方法进行查询。当然，在查询之前需要首先判断坐标变换矩阵是否为空，若不为空，需要对外包框矩形进行坐标变换后再执行查询，查询结果保存在要素集中，代码如下：

```
public void ExecuteIntersectionQuery(BoundingBox box, FeatureDataSet ds)
{
    if (CoordinateTransformation != null)
    {
        CoordinateTransformation.MathTransform.Invert();
        box = GeometryTransform.TransformBox(box,
            CoordinateTransformation.MathTransform);
        CoordinateTransformation.MathTransform.Invert();
```

```
_dataSource. Open( );
_dataSource. ExecuteIntersectionQuery( box, ds);
_dataSource. Close( );
}
```

4.5.2.5　Render（Graphics g, Map map）

这个方法是 SharpMap 中最重要的方法，是 SharpMap 整个绘图机制的核心。它分为两种情况：一种是当属性 Theme 不为空时，用 Theme 绘制地图；一种是当 Theme 为空时，用 Style 属性绘制地图。Theme 可提供更丰富的地图表达，关于 Theme 与 Style 的内容参见第 6 章。

用 Theme 绘制地图时，它通过 SharpMap. Data. FeatureDataTable 访问每一个 feature 进行绘制，先绘制其外框，再绘制内部几何体；用 Style 属性绘制地图时，它通过 DataSource. GetGeometriesInView（envelope）直接对得到几何体进行绘制，也是先绘制其外框，再绘制内部几何体。最后的绘制工作都调用 SharpMap. Rendering. VectorRenderer 类的函数完成。SharpMap. Rendering. VectorRenderer 类包含绘制注记、线串、复合线串、多点、复合多边形、点、多边形等。

绘制好图层后调用基类 Render 函数以触发 LayerRendered 事件，具体代码如下：

```
public override void Render( Graphics g, Map map)
{
```

首先判断地图中心点是否为空，若为空则抛出异常，提示不能渲染地图，地图中心不能明确。

```
if ( map. Center  = = null)
    throw ( new ApplicationException("Cannot render map. View center not
        specified")) ;
```

设置 g. SmoothingMode 属性为图层的 SmoothingMode 值，接着获取地图的外包框矩形，保存于 BoundingBox 类型的变量 envelope 中，若坐标变换矩阵不为空，则调用 GeometryTransform. TransformBox 对此外包框矩形进行坐标变换。

```
g. SmoothingMode = SmoothingMode;
BoundingBox envelope = map. Envelope;
if ( CoordinateTransformation !  = null)
{
    CoordinateTransformation. MathTransform. Invert( );
    envelope = GeometryTransform. TransformBox( envelope,
        CoordinateTransformation. MathTransform) ;
    CoordinateTransformation. MathTransform. Invert( );
}
```

若数据源为空，则抛出异常，提示未设置数据源属性。

```
if ( DataSource = = null)
```

4.5 矢量图层 VectorLayer

```
            throw (new ApplicationException("DataSource property not set
        on layer '" + LayerName + "'"));
        //If thematics is enabled, we use a slighty different rendering Approach
```

若 Theme 属性不为空，先实例化一个 FeatureDataSet 类型的对象 ds，用于保存相交查询的结果，打开数据源，调用数据源的 ExecuteIntersectionQuery 方法执行相交查询，然后关闭数据源。若坐标变换不为空，遍历查询结果中的每个要素，对几何对象属性做坐标变换处理。

```
        if (Theme ! = null)
        {
            FeatureDataSet ds = new FeatureDataSet();
            DataSource.Open();
            DataSource.ExecuteIntersectionQuery(envelope, ds);
            DataSource.Close();
            FeatureDataTable features = ds.Tables[0];
            if (CoordinateTransformation ! = null)
              for (int i = 0; i < features.Count; i + +)
                features[i].Geometry =
            GeometryTransform.TransformGeometry(features[i].Geometry,
            CoordinateTransformation.MathTransform);
```

若样式的 EnableOutline 属性值为 True，则绘制外边线。这里遍历每个查询结果，调用 Theme.GetStyle 方法计算要素数据行的样式。

```
        if (Style.EnableOutline)
        {
          for (int i = 0; i < features.Count; i + +)
        {
          FeatureDataRow feature = features[i];
          VectorStyle outlineStyle = Theme.GetStyle(feature) as VectorStyle;
          if (outlineStyle = = null) continue;
          if (! (outlineStyle.Enabled && outlineStyle.EnableOutline))
        continue;
          if (! (outlineStyle.MinVisible < = map.Zoom && map.Zoom < = outlineStyle.MaxVisible)) continue;
```

若几何对象属性为 LineString 类型，调用 VectorRenderer 类的 DrawLineString 方法绘制外边线；若为 MultiLineString 类型，调用 VectorRenderer 类的 DrawMultiLineString 方法绘制外边线。

```
            if (feature.Geometry is LineString)
            {
              VectorRenderer.DrawLineString(g, feature.Geometry as
              LineString, outlineStyle.Outline, map);
            }
```

```
            else if (feature. Geometry is MultiLineString)
            {
                VectorRenderer. DrawMultiLineString(g, feature. Geometry as
                    MultiLineString,outlineStyle. Outline, map);
            }
        }
    }
}
```

最后绘制要素内部。同样遍历每个要素，计算其样式，对计算的样式做合法性判断后，调用 RenderGeometry 方法绘制几何对象。

```
for (int i = 0; i < features. Count; i++)
{
    FeatureDataRow feature = features[i];
    VectorStyle style = Theme. GetStyle(feature) as VectorStyle;
    if (style = = null) continue;
    if (! style. Enabled) continue;
    if (!(style. MinVisible < = map. Zoom && map. Zoom < =
        style. MaxVisible)) continue;
        RenderGeometry(g, map, feature. Geometry, style);
}
```

若属性 Theme 为空，先判断图层的 Style. Enbled 属性，若为 false，则不需要渲染，直接返回。否则，打开数据源，调用数据源的 GetGeometriesInView 方法获取 envelope 内的几何对象，保存于 Collection < Geometry > 类型的变量 geoms 中，关闭数据源。若坐标变换属性不为空，对于每个几何对象做坐标变换后替换原来的几何对象。

```
else
{
    //if style is not enabled, we don't need to render anything
    if (! Style. Enabled) return;
    DataSource. Open();
    Collection < Geometry > geoms = ataSource. GetGeometriesInView(envelope);
    DataSource. Close();
    if (CoordinateTransformation ! = null)
        for (int i = 0; i < geoms. Count; i++)
            geoms[i] = GeometryTransform. TransformGeometry(geoms[i],
                CoordinateTransformation. MathTransform);
```

若样式的 EnableOutline 属性为真，遍历几何对象，绘制外边线。

```
    if (Style. EnableOutline)
    {
        foreach (Geometry geom in geoms)
        {
```

```
            if (geom != null)
            {
                //Draw background of all line-outlines first
                if (geom is LineString)
                    VectorRenderer.DrawLineString(g, geom as LineString, Style.Outline, map);
                else if (geom is MultiLineString)
                    VectorRenderer.DrawMultiLineString(g, geom as MultiLineString, Style.Outline, map);
            }
        }
```

遍历每个几何对象,调用 RenderGeometry 方法绘制每个几何对象,最后,调用基类的 Render 方法,完成绘制工作。

```
        for (int i = 0; i < geoms.Count; i++)
        {
            if (geoms[i] != null)
                RenderGeometry(g, map, geoms[i], Style);
        }
    }
    base.Render(g, map);
}
```

4.5.2.6 Dispose()

这个方法表示释放资源。这里直接调用数据源的 Dispose() 方法。代码如下:

```
public void Dispose()
{
    if (DataSource != null)
        DataSource.Dispose();
}
```

4.5.3 VectorLayer 的事件

VectorLayer 类中没有新定义事件,其事件就是基类 Layer 中定义的 LayerRendered 事件。

4.6 注记图层 LabelLayer

LabelLayer 用于在地图上显示地图注记。

4.6.1 LabelLayer 的属性

4.6.1.1 Envelope

这个属性和 VectorLayer 的属性一样,返回类型是 BoundingBox 对象,是通过调用数据源 GetExtents() 得到的。

4.6.1.2　DataSource

这个属性也和 VectorLayer 的一样。

4.6.1.3　LabelColumn

这个属性是获取和设置注记所使用数据源中的属性列名称，是字符串类型，代码如下：

```
private string _labelColumn;
public string LabelColumn
{
    get { return _labelColumn; }
    set { _labelColumn = value; }
}
```

4.6.1.4　LabelFilter

这个属性是设置和获取注记过滤器，是一个委托类型，用于过滤注记，如设置位置冲突检测，代码如下：

```
private LabelCollisionDetection.LabelFilterMethod _labelFilter;
public LabelCollisionDetection.LabelFilterMethod LabelFilter
{
    get { return _labelFilter; }
    set { _labelFilter = value; }
}
```

4.6.1.5　LabelStringDelegate

这个属性也是一个委托类型，它的作用是设置注记格式。例如，当我们需要将两个属性值"ROADNAME"和"STATE"结合到一起作为标注值时，可以给这个属性传递一个匿名委托，代码如下：

```
myLabelLayer.LabelStringDelegate = delegate(SharpMap.Data.FeatureDataRow fdr)
{ return fdr["ROADNAME"].ToString() + "," + fdr["STATE"].ToString(); };
```

这个属性的代码如下：

```
private GetLabelMethod _getLabelMethod;
public GetLabelMethod LabelStringDelegate
{
    get { return _getLabelMethod; }
    set { _getLabelMethod = value; }
}
```

4.6.1.6　MultipartGeometryBehaviour

若一个要素是由多要素组合而成的复合要素时，需设置这个属性，使用 MultipartGeometryBehaviourEnum 类型的枚举值，决定复合要素的绘制行为，代码如下：

```
private MultipartGeometryBehaviourEnum _multipartGeometryBehaviour;
```

4.6 注记图层 LabelLayer

```
public MultipartGeometryBehaviourEnum MultipartGeometryBehaviour
{
    get { return _multipartGeometryBehaviour; }
    set { _multipartGeometryBehaviour = value; }
}
```

4.6.1.7 Priority

这个属性值决定了注记在注记集合中的优先级，优先级大则显示在前面，优先级小则显示在后面，它是一个整形数据 int，代码如下：

```
private int _priority;
public int Priority
{
    get { return _priority; }
    set { _priority = value; }
}
```

4.6.1.8 PriorityColumn

这个属性是获取和设置注记优先级高的属性列，即数据源中用于注记的优先级最高的字段名称，是一个字符串类型，代码如下：

```
private string _priorityColumn = "";
public string PriorityColumn
{
    get { return _priorityColumn; }
    set { _priorityColumn = value; }
}
```

4.6.1.9 PriorityDelegate

这个属性也是委托类型，在计算渲染要素的优先级时，提供一些自定义的方法，若这个属性值不为空，将覆盖 PriorityColumn 的属性值。委托代码如下：

```
private GetPriorityMethod _getPriorityMethod;
public GetPriorityMethod PriorityDelegate
{
    get { return _getPriorityMethod; }
    set { _getPriorityMethod = value; }
}
```

4.6.1.10 RotationColumn

这个属性值决定了注记的旋转列，当这个值为空的时候，旋转的角度将是零或者是一个平行的线（旋转的单位是角度，方向为顺时针），代码如下：

```
private string _rotationColumn;
public string RotationColumn
{
```

```
get { return _rotationColumn; }
set { _rotationColumn = value; }
}
```

4.6.1.11 SmoothingMode

这个属性是获取和设置标注图层的表达质量,前面已经多次出现,不再赘述,代码如下:

```
private SmoothingMode _smoothingMode;
public SmoothingMode SmoothingMode
{
    get { return _smoothingMode; }
    set { _smoothingMode = value; }
}
```

4.6.1.12 SRID

这个属性是空间参考的 ID 编号,是一个可读写属性,在其 Set 访问器中,首先判断数据源是否为空,为空则抛出异常,提示数据源未设置;若不为空,返回数据源的空间参考的 ID 编号,代码如下:

```
public override int SRID
{
    get
    {
        if (DataSource == null)
            throw (new ApplicationException("DataSource property not set on layer '" + LayerName + "'"));
        return DataSource.SRID;
    }
    set { DataSource.SRID = value; }
}
```

4.6.1.13 Style

这个属性获取和设置注记图层的样式,这里使用的是 LabelStyle 类型,详细信息请参考第 6 章,代码如下:

```
public new LabelStyle Style
{
    get { return base.Style as LabelStyle; }
    set { base.Style = value; }
}
```

4.6.1.14 TextRenderingHint

这个属性获取和设置标注文本呈现模式,类型为枚举值 System.Drawing.Text.TextRenderingHint,代码如下:

```
private TextRenderingHint _textRenderingHint;
```

```
public TextRenderingHint TextRenderingHint
{
    get { return _textRenderingHint; }
    set { _textRenderingHint = value; }
}
```

4.6.1.15 Theme

这个属性获取和设置图层的主题，代码如下：

```
private ITheme _theme;
public ITheme Theme
{

    get { return _theme; }
    set { _theme = value; }
}
```

4.6.2 LabelLayer 的方法

4.6.2.1 构造函数 LabelLayer（string layername）

这个构造函数用给定的名称和默认的样式实例化一个注记图层，指定平滑效果为抗锯齿 SmoothingMode = SmoothingMode.AntiAlias，文本渲染特性为 TextRenderingHint.AntiAlias；复合要素（有多个要素组成的要素）的注记标注方法为全部标注，冲突检测方法为简单冲突检测，代码如下：

```
public LabelLayer(string layername)
    :base(new LabelStyle())
{
    LayerName = layername;
    SmoothingMode = SmoothingMode.AntiAlias;
    TextRenderingHint = TextRenderingHint.AntiAlias;
    _multipartGeometryBehaviour = MultipartGeometryBehaviourEnum.All;
    _labelFilter = LabelCollisionDetection.SimpleCollisionDetection;
}
```

4.6.2.2 Render（Graphics g, Map map）

这是标注图层的渲染方法，它针对 SharpMap.Data.FeatureDataTable 的每一个显示范围内的 feature 进行绘制，绘制工作都调用 SharpMap.Rendering.VectorRenderer 类 DrawLabel 函数完成，代码如下：

```
public override void Render(Graphics g, Map map)
{
```

先检查图层样式可用性，样式的最大可视化范围大于地图的缩放范围、样式的最小可视化范围小于地图缩放范围，都为真时，才继续操作。接着检查图层的数据源是否为空，若为空，抛出异常，提示数据源未设置。设置 Graphics 的文本呈现模式为标注图层的文本呈现模式，设置 Graphics 的呈现质量为图层的呈现质量属性值：

```
if ( Style. Enabled && Style. MaxVisible > = map. Zoom && Style. MinVisible < map. Zoom)
{
    if ( DataSource = = null)
        throw ( new ApplicationException ( "DataSource property not set on layer '" + LayerName
        + "'"));
    g. TextRenderingHint = TextRenderingHint;
    g. SmoothingMode = SmoothingMode;
```

接着调用地图的外包框矩形，保存在 BoundingBox 类型的变量 envelope 中，若坐标变换不为空，对此外包框矩形做坐标变换后，保存于变量 envelope 中：

```
BoundingBox envelope = map. Envelope; //View to render
if ( CoordinateTransformation ! = null)
{
    CoordinateTransformation. MathTransform. Invert( );
    envelope = GeometryTransform. TransformBox ( envelope, CoordinateTransformation. MathTransform);
    CoordinateTransformation. MathTransform. Invert( );
}
```

实例化一个要素数据集对象 ds 用于保存相交查询的结果，打开数据源，执行相交查询，关闭数据源。若查询的结果为空，调用基类的 Render 方法完成渲染并返回：

```
FeatureDataSet ds = new FeatureDataSet( );
DataSource. Open( );
DataSource. ExecuteIntersectionQuery( envelope, ds );
DataSource. Close( );
if ( ds. Tables. Count = = 0)
{
    base. Render(g, map);
    return;
}
```

将查询的要素数据表保存在变量 feature 中，实例化一个 Label 集合对象 labels，用于保存创建的标注对象。遍历每个要素，若坐标变换不为空，对每个要素的几何对象属性做坐标变换，并替换原来的几何对象：

```
FeatureDataTable features = ds. Tables[0];
List < Label > labels = new List < Label >( );
for ( int i = 0; i < features. Count; i + +)
{
    FeatureDataRow feature = features[i];
    if ( CoordinateTransformation ! = null)
        features[i]. Geometry = GeometryTransform. TransformGeometry(features[i]. Geometry,
```

4.6 注记图层 LabelLayer

CoordinateTransformation. MathTransform);

声明一个标注样式变量 style，若主题不为空，调用 Theme.GetStyle 方法计算要素的标注样式，保存于变量 style 中，否则直接将图层的样式赋予 style 变量：

```
LabelStyle style;
if (Theme ! = null) //If thematics is enabled, lets override the style
    style = Theme.GetStyle(feature) as LabelStyle;
else
    style = Style;
```

新建 float 变量 rotation，初始值为 0，表示标注的旋转角度为 0。若旋转列不为空，调用 feature 的索引方法获取旋转列值，调用 float.TryParse 转换为 float 类型后，保存于变量 rotation 中；整形变量 priority 用于保存标注的优先级，初始值为图层的优先级，若_getPriorityMethod 委托不为空，调用此委托计算优先级后，保存于变量 priority 中，若优先级列不为空，用要素的优先级值，转化为 int 类型后，保存于变量 priority 中；字符串型变量 text 用于保存标注的文本，若_getLabelMethod 委托不为空，调用此委托获取文本，保存于变量 text 中，否则，用要素的属性获取标注文本：

```
float rotation = 0;
if (! String.IsNullOrEmpty(RotationColumn))
    float.TryParse(feature[RotationColumn].ToString(), NumberStyles.Any,
Map.NumberFormatEnUs, out rotation);
int priority = Priority;
if (_getPriorityMethod ! = null)
    priority = _getPriorityMethod(feature);
else if (! String.IsNullOrEmpty(PriorityColumn))
    int.TryParse(feature[PriorityColumn].ToString(), NumberStyles.Any, Map. NumberForma-
        tEnUs, out priority);
string text;
if (_getLabelMethod ! = null)
    text = _getLabelMethod(feature);
else
    text = feature[LabelColumn].ToString();
```

若标注文本不为空，根据复合几何对象的类型，创建不同的标注对象；若复合几何对象的标注行为为将标注置于所有部分上，则遍历复合几何对象中的单个几何对象，调用 CreateLabel 方法为每个几何对象创建标注对象，参数为单个几何对象和前面计算的文本、旋转角度、优先级、样式、地图等，并将其加入标注对象集合中，代码如下：

```
if (text ! = null && text ! = String.Empty)
{
    if (feature.Geometry is GeometryCollection)
    {
```

```
if ( MultipartGeometryBehaviour = = MultipartGeometryBehaviourEnum. All)
{
    foreach ( Geometry geom in ( feature. Geometry as GeometryCollection) )
    {
        Label lbl = CreateLabel(geom, text, rotation, priority, style, map, g);
        if ( lbl ! = null) labels. Add(lbl);
    }
}
```

若复合几何对象的标注行为为将标注置于几何中心,则调用 CreateLabel 方法,创建标注对象,不同的是,这里直接使用要素的 Geometry 属性作为第一个参数:

```
else if ( MultipartGeometryBehaviour = = MultipartGeometryBehaviourEnum. CommonCenter)
{
    Label lbl = CreateLabel(feature. Geometry, text, rotation, priority, style, map, g);
    if ( lbl ! = null) labels. Add( lbl);
}
```

若复合几何对象的标注行为为将标注置于第一个单个几何对象上,则调用 CreateLabel 方法创建标注对象,这里首先判断要素是否为复合要素,若是则将第一个单个几何对象作为 CreateLabel 方法的第一个参数,代码如下:

```
else if ( MultipartGeometryBehaviour = = MultipartGeometryBehaviourEnum. First)
{
    if ( ( ( feature. Geometry as GeometryCollection). Collection. Count > 0)
    {
        Label lbl = CreateLabel( ( feature. Geometry as GeometryCollection). Collection[0], text,
        rotation, style, map, g);
        if ( lbl ! = null) labels. Add(lbl);
    }
}
```

若复合几何对象的标注行为为将标注置于最大的单个几何对象上,则根据几何对象的类型,计算最大几何对象:首先将要素的几何对象转换为 GeometryCollection 类型,保存于变量 coll 中,若 coll 的几何对象个数属性 NumGeometries 的值大于 0,则继续操作。遍历 coll 中的每个几何对象,若几何对象的类型为 LineString,则几何对象的长度属性 Length 值大的作为最大几何对象;若几何对象的类型为 MultiLineString,则将几何对象强制转换为 MultiLineString 类型,以几何对象长度属性 Length 值最大的作为最大几何对象;若几何对象的类型为 Polygon,则以面积最大的作为最大几何对象;若几何对象的类型为 MultiPolygon,则将其强制转换为 MultiPolygon 类型,以面积最大的作为最大几何对象;最后调用 CreateLabel 方法创建标注对象,第一个参数为刚查找的最大几何对象:

```
else if ( MultipartGeometryBehaviour = = MultipartGeometryBehaviourEnum. Largest)
{
    GeometryCollection coll = ( feature. Geometry as GeometryCollection);
```

4.6 注记图层 LabelLayer

```
            if (coll.NumGeometries > 0)
            {
                double largestVal = 0;
                int idxOfLargest = 0;
                for (int j = 0; j < coll.NumGeometries; j++)
                {
                    Geometry geom = coll.Geometry(j);
                    if (geom is LineString && ((LineString) geom).Length > largestVal)
                    {
                        largestVal = ((LineString) geom).Length;
                        idxOfLargest = j;
                    }
                    if (geom is MultiLineString && ((MultiLineString) geom).Length > largestVal)
                    {
                        largestVal = ((MultiLineString)geom).Length;
                        idxOfLargest = j;
                    }
                    if (geom is Polygon && ((Polygon) geom).Area > largestVal)
                    {
                        largestVal = ((Polygon) geom).Area;
                        idxOfLargest = j;
                    }
                    if (geom is MultiPolygon && ((MultiPolygon) geom).Area > largestVal)
                    {
                        largestVal = ((MultiPolygon) geom).Area;
                        idxOfLargest = j;
                    }
                }
Label lbl = CreateLabel(coll.Geometry(idxOfLargest), text, rotation, priority, style, map, g);
                if (lbl! = null) labels.Add(lbl);
            }
        }
    }
    else
    {
        Label lbl = CreateLabel(feature.Geometry, text, rotation, priority, style, map, g);
        if (lbl! = null) labels.Add(lbl);
    }
}
```

若生成的标注对象的个数大于0，则渲染标注对象。首先，若样式的CollisionDetection属性为真，且_labelFilter不为空，则调用_labelFilter委托过滤标注，否则，遍历每个标注对象，若标注的 Show 属性值为 True，调用 VectorRenderer 类的 DrawLabel 静态方法，渲染标注对象。方法的最后调用基类 Render 方法完成渲染。

```
if (labels.Count > 0) //We have labels to render...
{
if (Style.CollisionDetection && _labelFilter != null) _labelFilter(labels);
for (int i = 0; i < labels.Count; i++)
  if (labels[i].Show) VectorRenderer.DrawLabel(g, labels[i].LabelPoint, labels[i].Style.Offset,
labels[i].Style.Font, labels[i].Style.ForeColor, labels[i].Style.BackColor, Style.Halo, labels
[i].Rotation, labels[i].Text, map);
}
}
base.Render(g, map);
}
```

4.6.2.3 CreateLabel（Geometry feature, string text, float rotation, int priority, LabelStyle style, Map map, Graphics g）

此方法创建一个标注对象，参数为要创建标注的几何对象 feature、标注的文本 text、标注的旋转角度 rotation、标注的优先级 priority、标注的样式 style、标注的地图 map、Graphics 对象 g。首先调用 VectorRenderer.SizeOfString 方法计算标注文本的大小，然后计算标注的位置，若样式的 CollisionDetection 属性值为 false，将构造函数的第五个参数设置为 null，直接创建标注对象；否则，新生成一个 LabelBox 对象，作为构造函数的第五个参数，生成标注对象，代码如下：

```
private static Label CreateLabel(Geometry feature, string text, float rotation, int priority, LabelStyle style, Map map, Graphics g)
{
SizeF size = VectorRenderer.SizeOfString(g, text, style.Font);
PointF position = map.WorldToImage(feature.GetBoundingBox().GetCentroid());
position.X = position.X - size.Width * (short)style.HorizontalAlignment * 0.5f;
position.Y = position.Y - size.Height * (short)(2 - (int)style.VerticalAlignment) * 0.5f;
if (position.X - size.Width > map.Size.Width || position.X + size.Width < 0 ||
   position.Y - size.Height > map.Size.Height || position.Y + size.Height < 0)
   return null;

Label lbl;
if (!style.CollisionDetection)
   lbl = new Label(text, position, rotation, priority, null, style);
else
{
    //Collision detection is enabled so we need to measure the size of the string
    lbl = new Label(text, position, rotation, priority, new LabelBox(position.X - size.Width *
0.5f - style.CollisionBuffer.Width, position.Y + size.Height * 0.5f + style.CollisionBuffer.
```

4.6 注记图层 LabelLayer

```
            Height, size.Width + 2f * style.CollisionBuffer.Width, size.Height + style.CollisionBuffer.Height
        * 2f), style);
    }
    if (feature is LineString)
    {
        LineString line = feature as LineString;
        if (line.Length/map.PixelSize > size.Width) //Only label feature if it is long enough
            CalculateLabelOnLinestring(line, ref lbl, map);
        else return null;
    }
    return lbl;
}
```

4.6.2.4 CreateLabel (Geometry feature, string text, float rotation, LabelStyle style, Map map, Graphics g)

这是创建注记的一个重载方法,比上一个方法少了 Priority 参数,它使用图层的 Priority 作为参数,在内部仍然调用上一个函数进行注记的创建工作,代码如下:

```
private Label CreateLabel(Geometry feature, string text, float rotation, LabelStyle style, Map map,
    Graphics g)
{
    return CreateLabel(feature, text, rotation, Priority, style, map, g);
}
```

4.6.2.5 CalculateLabelOnLinestring (LineString line, ref Label label, Map map)

此方法对字符线串进行调整,计算标注的位置。这里先计算中间点的索引,保存于变量 midPoint 中。若线的节点数大于 2,计算中间点下一点和中间点的 X、Y 距离,保存于 double 类型的变量 dx、dy 中;若线的节点数不大于 2,则设置 midPoint 值为 0,计算第二点和第一点的距离,保存于 dx、dy 中,代码如下:

```
private static void CalculateLabelOnLinestring(LineString line, ref Label label, Map map)
{
    double dx, dy;
    int midPoint = (line.Vertices.Count - 1)/2;
    if (line.Vertices.Count > 2)
    {
        dx = line.Vertices[midPoint + 1].X - line.Vertices[midPoint].X;
        dy = line.Vertices[midPoint + 1].Y - line.Vertices[midPoint].Y;
    }
    else
    {
        midPoint = 0;
        dx = line.Vertices[1].X - line.Vertices[0].X;
        dy = line.Vertices[1].Y - line.Vertices[0].Y;
```

若 dy 值为零，设置标注的旋转角度为 0；若 dx 值为 0，设置标注的旋转角度为 90；其他情况，先计算线的角度，然后再计算标注的旋转角度，最后计算标注的坐标点，设置注记对象的 LabelPoint 属性：

```
if ( dy = = 0 )
    label. Rotation = 0;
else if ( dx = = 0 )
    label. Rotation = 90;
else
{
    // calculate angle of line
    double angle = - Math. Atan( dy/dx) + Math. PI * 0.5;
    angle * = (180d/Math. PI); // convert radians to degrees
    label. Rotation = (float) angle - 90; // -90 text orientation
}
double tmpx = line. Vertices[midPoint]. X + (dx * 0.5);
double tmpy = line. Vertices[midPoint]. Y + (dy * 0.5);
label. LabelPoint = map. WorldToImage(new Point(tmpx, tmpy));
}
```

4.6.2.6　Dispose ( )

这里主要调用数据源的 Dispose ( ) 方法进行资源的释放，代码如下：

```
public void Dispose( )
{
    if ( DataSource ! = null) DataSource. Dispose( );
}
```

4.7　Layer 集合

LayerCollection 是 Layer 对象的集合，派生自 CollectionBase。

4.7.1　LayerCollection 的索引器

4.7.1.1　this [int index]

图层的索引器。它实现了 C#的索引器模式，LayerCollection 类中的 List 实际上是继承自 CollectionBase 的属性，详细代码如下：

```
public virtual ILayer this[int index]
{
    get { return (ILayer) List[index]; }
    set { List[index] = value; }
}
```

4.7.1.2 this [string layerName]

通过名称来进行索引的索引器,它调用 GetLayerByName 方法得到对应图层,在 Set 访问器中,若与现有的图层名称相同,则更新图层;若不同则调用 Add() 方法添加图层,代码如下:

```
public virtual ILayer this[string layerName]
{
    get { return GetLayerByName(layerName); }
    set
    {
        for (int i = 0; i < Count; i++)
        {
            int comparison = String.Compare(this[i].LayerName,
            layerName, StringComparison.CurrentCultureIgnoreCase);
            if (comparison == 0)
            {
                this[i] = value;
                return;
            }
        }
        Add(value);
    }
}
```

4.7.2 LayerCollection 的方法

4.7.2.1 Insert (int index, ILayer layer)

在指定的位置插入图层,调用的是 List 的 insert 方法,代码如下:

```
public void Insert(int index, ILayer layer)
{
    if (index >= Count || index < 0)
    {
        throw new ArgumentOutOfRangeException("index", index, "Index not in range");
    }
    List.Insert(index, layer);
}
```

4.7.2.2 IndexOf (ILayer layer)

通过图层得到图层的索引号,调用 List 的 IndexOf 方法进行包装,代码如下:

```
public int IndexOf(ILayer layer)
{
    return List.IndexOf(layer);
}
```

4.7.2.3 Add（ILayer layer）

添加图层，实际上调用 List 的 Add 方法进行包装，代码如下：

```
public void Add(ILayer layer)
{
    List.Add(layer);
}
```

4.7.2.4 GetLayerByName（string layerName）

通过图层的名称得到图层，它遍历每个图层，对图层的名称和给定的名称进行对比，若相等则返回这个图层，代码如下：

```
private ILayer GetLayerByName(string layerName)
{
    foreach (ILayer layer in this)
    {
        int comparison = String.Compare(layer.LayerName,
                                        layerName,
                                        StringComparison.CurrentCultureIgnoreCase);
        if (comparison == 0) return layer;
    }
    return null;
}
```

4.7.2.5 Remove（ILayer layer）

移除图层，它是对 List 的 remove 方法进行包装，若传进的参数为空，则没有任何效果，代码如下：

```
public void Remove(ILayer layer)
{
    List.Remove(layer);
}
```

4.7.2.6 RemoveAt（int index）

通过指定的索引移除图层，这是对 List 的 RemoveAt 方法进行包装，代码如下：

```
public new void RemoveAt(int index)
{
    List.RemoveAt(index);
}
```

复习思考题

4-1 SharpMap 中实现 ILayer 接口的图层类有哪些，ILayer 的主要功能是什么？

4-2 SharpMap 中图层管理的方式是什么？试编程实现具有图层添加、删除、移动、查找等功能。

第5章 绘 制

SharpMap 中有一个 SharpMap.Rendering（绘制）的名称空间，其下包含了各种与图形绘制相关的对象。在第4章中我们已经看到，VectorLayer 与 LabelLayer 最终的绘制工作都是交给 VectorRender 的相关静态函数完成的。SharpMap 中各种 Render 最终调用的 .NET 下 System.Drawing 中 GDI+ 库的绘制方法，完成地图的绘制。SharpMap.Rendering 名称空间的类图如图 5-1 所示。

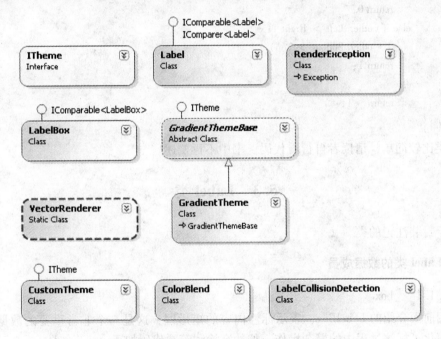

图 5-1　SharpMap.Rendering 名称空间的类图

5.1　ClipState 枚举

这个定义很简单，定义了内部、外部、相交三种裁减状态。代码如下：

```
private enum ClipState
{
    Within,
    Outside,
    Intersecting
}
```

5.2 LabelBox 类

LabelBox 类主要提供了对标注外框的封装，实际上就是一个矩形。LabelBox 有如下两个函数：

（1）Intersects（LabelBox box），判断两个矩形框是否相交，如相交则返回真。

（2）CompareTo（LabelBox other），比较两个矩形框。两个矩形相交时，返回 0，表示相等；否则左边或下边的矩阵被认为较大，返回 1。代码如下：

```
public int CompareTo(LabelBox other)
{
    if (Intersects(other))
        return 0;
    else if (other.Left > Right ||
        other.Bottom > Top)
        return 1;
    else
        return -1;
}
```

该类比较简单，请读者自行看代码，书中不再赘述。

5.3 Label 类

用于封装注记的类。

5.3.1 Label 类的数据成员

5.3.1.1 _box

使用前面介绍的 LabelBox，表示标注内容的矩形框，这是一个非常重要的数据成员，它大大简化了标注过程中的繁杂操作，如冲突检测等，代码如下：

```
private LabelBox _box;
```

5.3.1.2 _Font

绘制标注时所使用的字体，它直接使用 .NET 的 font 类，代码如下：

```
private Font _Font;
```

5.3.1.3 _LabelPoint

绘制标注的坐标点，它是一个 PointF 结构的值，代码如下：

```
private PointF _LabelPoint;
```

5.3.1.4 _Priority

绘制标注的优先级，在冲突检测时非常有用，代码如下：

```
private int _Priority;
```

5.3.1.5 _Rotation

绘制标注旋转的角度,是一个 float 类型的数值,代码如下:

```
private float _Rotation;
```

5.3.1.6 _show

这是一个布尔类型的标记值,值为 false 的标注在绘制时将不显示,代码如下:

```
private bool _show;
```

5.3.1.7 _Style

绘制标注时所使用的样式,它使用了 SharpMap 的 LabelStyle 类型,详细信息请参考第 6 章,代码如下:

```
private LabelStyle _Style;
```

5.3.1.8 _Text

绘制标注时的文本内容,是一个字符串类型的值,代码如下:

```
private string _Text;
```

5.3.2 Label 类的属性

Label 类的属性分别为 Text、Show、Box、labelPoint、Font、Priority、Rotation、Style、Text,只是简单的对数据成员进行了封装,这里不再赘述。

5.3.3 Label 类的方法

5.3.3.1 构造函数

这个构造函数使用标注的文本、坐标点、旋转值、优先级、外包矩形框、样式来生成一个标注,代码如下:

```
public Label(string text, PointF labelpoint, float rotation, int priority, LabelBox collisionbox, LabelStyle style)
{
    _Text = text;
    _LabelPoint = labelpoint;
    _Rotation = rotation;
    _Priority = priority;
    _box = collisionbox;
    _Style = style;
    _show = true;
}
```

5.3.3.2 CompareTo(Label other)

这是比较方法。这里将标注与传进来的标注的外包矩形比较,使用 LabelBox 类的

CompareTo 方法，代码如下：

```csharp
public int CompareTo(Label other)
{
    if (this == other)
        return 0;
    else if (_box == null)
        return -1;
    else if (other.Box == null)
        return 1;
    else
        return _box.CompareTo(other.Box);
}
```

5.3.3.3　Compare (Label x, Label y)

这是 Compare 的一个重载方法，在内部直接调用 CompareTo 方法，代码如下：

```csharp
public int Compare(Label x, Label y)
{
    return x.CompareTo(y);
}
```

5.4　LabelCollisionDetection 类

这个类主要用于标注的冲突检测及处理。

5.4.1　LabelCollisionDetection 类的方法

LabelCollisionDetection 类内的两个函数均为静态函数。

5.4.1.1　SimpleCollisionDetection (List<Label> labels)

这个方法是一个简单而快速的标注检测方法，先对 Label 的集合进行排序，然后进行一次遍历，对相邻 Label 进行冲突检测，如有冲突情况，移除相交时优先级小的标注，代码如下：

```csharp
public static void SimpleCollisionDetection(List<Label> labels)
{
    labels.Sort();
    for (int i = labels.Count - 1; i > 0; i--)
        if (labels[i].CompareTo(labels[i - 1]) == 0)
        {
            if (labels[i].Priority == labels[i - 1].Priority) continue;
            if (labels[i].Priority > labels[i - 1].Priority) labels.RemoveAt(i - 1);
            else labels.RemoveAt(i);
        }
}
```

5.4.1.2 ThoroughCollisionDetection（List < Label > labels）

这是全面而详细的冲突检测方法，这个方法对每个标注都进行比较，根据优先级，将有冲突的优先级小的标注的 Label 的 Show 属性设置为 False，可以看到，这个方法虽然比较时很全面，但算法的复杂度也大为上升，代码如下：

```csharp
public static void ThoroughCollisionDetection(List < Label > labels)
{
    labels.Sort();
    for (int i = labels.Count - 1; i > 0; i--)
    {
        if (!labels[i].Show) continue;
        for (int j = i - 1; j >= 0; j--)
        {
            if (!labels[j].Show) continue;
            if (labels[i].CompareTo(labels[j]) == 0)
            if (labels[i].Priority >= labels[j].Priority) {labels[j].Show = false;}
            else {labels[i].Show = false; break;}
        }
    }
}
```

5.4.2 LabelCollisionDetection 类的代理

相关代码如下：

```csharp
public delegate void LabelFilterMethod(List < Label > labels);
```

5.5 矢量数据渲染类 VectorRender

VectorRender 矢量数据渲染类负责对矢量数据进行绘制。

5.5.1 VectorRender 类的数据成员

5.5.1.1 ExtremeValueLimit

极值的限制，超过该极限值的数被系统认为是非法值，代码如下：

```csharp
private const float ExtremeValueLimit = 1E + 8f;
```

5.5.1.2 NearZero

零的近似值，由于计算过程中的精度问题，所以小于该值的数将被认为是 0，代码如下：

```csharp
private const float NearZero = 1E - 30f; // 1/Infinity
```

5.5.1.3 Defaultsymbol

默认的点符号,可以看出,sharpmap 中的点符号是使用的栅格图片绘制的,默认值是一个系统自带的图片 DefaultSymbol.png,代码如下:

```
private static readonly Bitmap Defaultsymbol = (Bitmap) Image.FromStream(Assembly.GetExecutingAssembly().GetManifestResourceStream("SharpMap.Styles.DefaultSymbol.png"));
```

5.5.2 VectorRender 类的属性

SizeOfString 是一个委托类型,允许用户自定义计算字符显示宽度的方法,定义如下:

```
public delegate SizeF SizeOfStringDelegate(Graphics g, string text, Font font);
private static SizeOfStringDelegate _sizeOfString;
public static SizeOfStringDelegate SizeOfString
{
    get { return _sizeOfString; }
    set
    {
        if (value ! = null)
            _sizeOfString = value;
    }
}
```

5.5.3 VectorRender 类的方法

5.5.3.1 DrawLabel

此方法在地图上绘制一个标注,参数为 Graphics 对象、注记坐标 LabelPoint、注记偏移量 Offset、注记的字体 font、注记的前景色 forecolor、注记的背景色 backcolor、注记的光环 halo、注记的旋转角度 rotation、注记的文本 text、标注的地图对象 map,代码如下:

```
public static void DrawLabel(System.Drawing.Graphics g, System.Drawing.PointF LabelPoint, System.Drawing.PointF Offset, System.Drawing.Font font, System.Drawing.Color forecolor, System.Drawing.Brush backcolor, System.Drawing.Pen halo, float rotation, string text, SharpMap.Map map)
{
```

首先调用_sizeOfString 委托,计算注记文本的大小,然后将标注的坐标加上偏移值,赋予标注坐标 labelPoint;若旋转角度不为零且不为空,调用 g 的 TranslateTransform 方法改变图形对象的原点,平移的值为 labelPoint 坐标值;调用 g 的 RotateTransform 方法进行旋转,然后再调用 g 的 TranslateTransform 方法改变图形对象的原点,平移大小为标注字体大小的一半:

```
SizeF fontSize = _sizeOfString(g, text, font); //Calculate the size of the text
labelPoint.X + = offset.X;
```

```
labelPoint.Y + = offset.Y; //add label offset
if (rotation ! = 0 && ! float.IsNaN(rotation))
{
    g.TranslateTransform(labelPoint.X, labelPoint.Y);
    g.RotateTransform(rotation);
    g.TranslateTransform( - fontSize.Width/2, - fontSize.Height/2);
```

若背景色不为空，且背景色不为透明色，调用 g 的 FillRetangle 方法绘制背景框。新建一个 GraphicsPath 对象 path，调用其 AddString 方法添加要绘制的文字——标注文本；若光环 halo 不为空，调用 d.DrawPath 绘制标注文本，设置 g.TransForm 值为地图的 MapTransform 属性值：

```
if (backcolor ! = null && backcolor ! = Brushes.Transparent)
    g.FillRectangle(backcolor, 0, 0, fontSize.Width * 0.74f + 1f, fontSize.Height * 0.74f);
GraphicsPath path = new GraphicsPath();
path.AddString(text, font.FontFamily, (int) font.Style, font.Size, new System.Drawing.Point(0, 0), null);
if (halo ! = null) g.DrawPath(halo, path);
g.FillPath(new SolidBrush(forecolor), path);
g.Transform = map.MapTransform;
}
```

若地图旋转为空，则直接绘制标注。首先，若背景色不为空，且不为透明色，调用 g.FillRectangle 绘制背景框，绘制文本过程与前面一样，代码如下：

```
else
{
    if (backcolor ! = null && backcolor ! = Brushes.Transparent)
        g.FillRectangle(backcolor, labelPoint.X, labelPoint.Y, fontSize.Width * 0.74f + 1, fontSize.Height * 0.74f);
    GraphicsPath path = new GraphicsPath();
    path.AddString(text, font.FontFamily, (int) font.Style, font.Size, labelPoint, null);
    if (halo ! = null) g.DrawPath(halo, path);
    g.FillPath(new SolidBrush(forecolor), path);
}
```

5.5.3.2 DrawLineString

此方法在地图上绘制一个字符线串，首先判断线的节点数是否大于1，若大于1，新建一个 GraphicsPath 对象 gp，用于保存要绘制的线条。调用 gp 的 AddLines 方法，添加要绘制的线条，最后调用 Graphics 的 DrawPath 绘制线条，代码如下：

```
public static void DrawLineString(Graphics g, LineString line, Pen pen, Map map)
{
    if (line.Vertices.Count > 1)
    {
```

```
            GraphicsPath gp = new GraphicsPath();
            gp.AddLines(LimitValues(line.TransformToImage(map), ExtremeValueLimit));
            g.DrawPath(pen, gp);
        }
    }
```

5.5.3.3　DrawMultiLineString

此方法绘制复合线串对象，它通过循环调用 DrawLineString 进行绘制，代码如下：

```
    public static void DrawMultiLineString(Graphics g, MultiLineString lines, Pen pen, Map map)
    {
        for (int i = 0; i < lines.LineStrings.Count; i++)
            DrawLineString(g, lines.LineStrings[i], pen, map);
    }
```

5.5.3.4　DrawMultiPoint

此方法绘制复合点对象，它通过循环调用 Graphics 类的 DrawPoint 进行绘制，代码如下：

```
    public static void DrawMultiPoint(Graphics g, MultiPoint points, Image symbol, float symbolscale,
        PointF offset, float rotation, Map map)
    {
        for (int i = 0; i < points.Points.Count; i++)
            DrawPoint(g, points.Points[i], symbol, symbolscale, offset, rotation, map);
    }
```

5.5.3.5　DrawMultiPolygon

此方法绘制复合多边形对象，它通过循环调用 Graphics 类的 DrawPolygon 进行绘制，代码如下：

```
    public static void DrawMultiPolygon(Graphics g, MultiPolygon pols, Brush brush, Pen pen, bool
        clip, Map map)
    {
        for (int i = 0; i < pols.Polygons.Count; i++)
            DrawPolygon(g, pols.Polygons[i], brush, pen, clip, map);
    }
```

5.5.3.6　DrawPoint

此方法绘制点对象，参数中，symbol 参数为一个位图，表示点的符号。最终绘制调用 Graphics 类的 DrawImage。若标注点为空，则直接返回；若符号为空，则直接返回，代码如下：

```
    public static void DrawPoint(Graphics g, Point point, Image symbol, float symbolscale, PointF offset,
        float rotation, Map map)
    {
        if (point == null) return;
```

5.5 矢量数据渲染类 VectorRender

```
if ( symbol = = null ) symbol = Defaultsymbol;
```

调用 Transform 类的 WorldtoMap 方法,将标注点的坐标转换为地图坐标,保存在点 pp 中;保存 g.Transform 对象,以便使用后还原:

```
PointF pp = Transform.WorldtoMap( point, map);
Matrix startingTransform = g.Transform;
```

若旋转角度不为 0,且不为空,计算旋转中心,保存于点 rotationCenter 中;然后新建一个 Martix 对象 transformation,用于旋转变换,然后调用 transform 的 RotateAt 方法,进行旋转操作,并将新的矩阵赋予 g.Transform 属性;若符号的比例范围为 1,则调用 g.DrawImageUnscaled 方法,绘制出原始大小的符号;否则,计算出需要绘制的符号的高度和宽度,调用 g.DrawImage 方法绘制出符号:

```
if ( rotation ! = 0 && ! Single.IsNaN( rotation ) )
    {
    PointF rotationCenter = PointF.Add( pp, new SizeF( symbol.Width/2, symbol.Height/2 ) );
    Matrix transform = new Matrix( );
    transform.RotateAt( rotation, rotationCenter );
    g.Transform = transform;
    if ( symbolscale = = 1f )
        g.DrawImageUnscaled( symbol, ( int ) ( pp.X – symbol.Width/2 + offset.X ),( int ) ( pp.Y
          – symbol.Height/2 + offset.Y ) );
    else
        {
        float width = symbol.Width * symbolscale;
        float height = symbol.Height * symbolscale;
        g.DrawImage( symbol, ( int ) pp.X – width/2 + offset.X * symbolscale,
                    ( int ) pp.Y – height/2 + offset.Y * symbolscale, width, height );
        }
    g.Transform = startingTransform;
    }
```

若旋转角度为 0 或者为空,则不需要进行旋转变换,直接根据符号的比例绘制符号,过程与上面相同,请参考以上解释:

```
else
    {
    if ( symbolscale = = 1f )
      g.DrawImageUnscaled( symbol, ( int ) ( pp.X – symbol.Width/2 + offset.X ),
                          ( int ) ( pp.Y – symbol.Height/2 + offset.Y ) );
    else
        {
        float width = symbol.Width * symbolscale;
        float height = symbol.Height * symbolscale;
```

```
    g. DrawImage( symbol, ( int) pp. X  -  width/2  +  offset. X * symbolscale,
                  ( int) pp. Y  -  height/2  +  offset. Y * symbolscale, width, height) ;
   }
  }
 }
```

5.5.3.7 DrawPolygon

此为绘制多边形要素的方法。参数为 Graphics 对象 g、要绘制的多边形对象 pol、画刷对象 brush、画笔对象 pen、是否裁减 clip、绘制的地图对象 map。它支持用不同的画刷和画笔来绘制多边形，也是采用 Graphics 类的 DrawPath 进行具体绘制。绘制过程分为两步：外环绘制、内环绘制。绘制外环时，首先检查外环多边形是否成环，若不成环则直接返回；接着判断多边形的节点数是否大于 2，若大于才继续操作，代码如下：

```
public static void DrawPolygon( Graphics g, Polygon pol, Brush brush, Pen pen, bool clip, Map
map)
{
    if ( pol. ExteriorRing  = =  null)
       return;
    if ( pol. ExteriorRing. Vertices. Count  >  2)
    {
        //Use a graphics path instead of DrawPolygon. DrawPolygon has a problem with several interior holes
```

新建一个 GraphPath 对象，用于绘制多边形（并没有使用 DrawPolygon，应为 DrawPolygon 方法本身有一些小漏洞）。若不需要裁减，调用 LimitValues 方法，将要绘制的多边形边界确定后，直接加入 GraphicsPath 中；若需要裁减，调用 DrawPolygonClipped 方法进行绘制：

```
GraphicsPath gp  =  new GraphicsPath( );
    if ( ! clip)
 gp. AddPolygon ( LimitValues ( pol. ExteriorRing. TransformToImage ( map ), ExtremeValueLimit) ) ;
        else
DrawPolygonClipped  ( gp,  pol. ExteriorRing. TransformToImage  ( map ),  map. Size. Width,
map. Size. Height) ;
```

绘制内环时，循环遍历每个内环，若需要裁减，调用 LimitValues 确定绘制边界后，直接加入 GraphicsPath 中；若不需要裁减，调用 DrawPolygonClipped 方法进行绘制：

```
for ( int i = 0; i < pol. InteriorRings. Count; i + + )
   if ( ! clip) gp. AddPolygon( LimitValues( pol. InteriorRings[ i ]. TransformToImage( map), ExtremeValueLimit) ) ;
   else DrawPolygonClipped ( gp, pol. InteriorRings [ i ]. TransformToImage ( map), map. Size. Width,
map. Size. Height) ;
```

若画笔不为空，且画笔不为透明色，调用 Graphics 的 FillPath 方法填充多边形；若画笔不为空，则调用 Graphics 的 DrawPath 方法绘制多边形外边框：

```
if (brush ! = null && brush ! = Brushes.Transparent) g.FillPath(brush, gp);
if (pen ! = null) g.DrawPath(pen, gp);
}
}
```

5.5.3.8　DrawPolygonClipped

此方法绘制裁剪后的多边形。这里首先调用 DetermineClipState 方法判断裁减的类型，若裁减类型为内裁减 ClipState.Within，则直接将多边形加入 GraphicsPath 中；若裁减类型为相交裁减 ClipState.Intersecting，调用 clipPolygon 方法进行裁减后，加入 GraphicsPath 中，代码如下：

```
private static void DrawPolygonClipped(GraphicsPath gp, PointF[] polygon, int width, int height)
{
    ClipState clipState = DetermineClipState(polygon, width, height);
    if (clipState == ClipState.Within)
    {
        gp.AddPolygon(polygon);
    }
    else if (clipState == ClipState.Intersecting)
    {
        PointF[] clippedPolygon = clipPolygon(polygon, width, height);
        if (clippedPolygon.Length > 2)
            gp.AddPolygon(clippedPolygon);
    }
}
```

5.5.3.9　LimitValues (PointF [ ] vertices, float limit)

此方法确定绘制边界。这个方法的目的就是为了防止在调用 FillPath 方法时发生"值溢出"异常（在坐标值过大（小）时），超过极值的点将会被设置为允许的最大（小）值，代码如下：

```
private static PointF[] LimitValues(PointF[] vertices, float)
{
    for (int i = 0; i < vertices.Length; i++)
    {
        vertices[i].X = Math.Max(-limit, Math.Min(limit, vertices[i].X));
        vertices[i].Y = Math.Max(-limit, Math.Min(limit, vertices[i].Y));
    }
    return vertices;
}
```

5.5.3.10　SizeOfStringBase (Graphics g, string text, Font font)

计算用指定的 Font 绘制时，指定字符串的宽度。实际上调用 Graphics.MeasureString

方法，代码如下：

```
public static SizeF SizeOfStringBase(Graphics g, string text, Font font)
{
    return g.MeasureString(text, font);
}
```

5.5.3.11　SizeOfStringCeiling（Graphics g，string text，Font font）

此方法对字符串的宽度进行取整操作，返回大于或等于指定数字的最小整数，代码如下：

```
public static SizeF SizeOfStringCeiling(Graphics g, string text, Font font)
{
    SizeF f = g.MeasureString(text, font);
    return new SizeF((float)Math.Ceiling(f.Width),(float)Math.Ceiling(f.Height));
}
```

5.5.3.12　DetermineClipState（PointF [ ] vertices，int width，int height）

此方法判断裁减类型，针对坐标点，先找到这个坐标点所在的边界点，如果在指定宽度和高度的矩形内部，则裁减状态为ClipState.Within；如果在外部则为ClipState.Outside；其他情况则返回ClipState.Intersecting，代码如下：

```
private static ClipState DetermineClipState(PointF[ ] vertices, int width, int height)
{
    float minX = float.MaxValue;
    float minY = float.MaxValue;
    float maxX = float.MinValue;
    float maxY = float.MinValue;
    for (int i = 0; i < vertices.Length; i++)
    {
        minX = Math.Min(minX, vertices[i].X);
        minY = Math.Min(minY, vertices[i].Y);
        maxX = Math.Max(maxX, vertices[i].X);
        maxY = Math.Max(maxY, vertices[i].Y);
    }
    if (maxX < 0) return ClipState.Outside;
    if (maxY < 0) return ClipState.Outside;
    if (minX > width) return ClipState.Outside;
    if (minY > height) return ClipState.Outside;
    if (minX > 0 && maxX < width && minY > 0 && maxY < height)
        return ClipState.Within;
    return ClipState.Intersecting;
}
```

5.5.3.13　clipPolygon（PointF [ ] vertices，int width，int height）

绘制多边形的过程中，若需要裁减，则会调用此方法裁减多边形。它会在给定的点集

5.5 矢量数据渲染类 VectorRender

合中选取落在指定高度和宽度矩形内部的点，以确定合适的绘制边界，达到"裁减"的效果。参数为多边形点的集合 vertics、图像的宽度 width、图像的高度 height。首先判断多边形点的个数是否小于或等于 1，若是则不需要裁减，直接返回方法，代码如下：

```
internal static PointF[ ] clipPolygon( PointF[ ] vertices, int width, int height)
{
    float deltax, deltay, xin, xout, yin, yout;
    float tinx, tiny, toutx, touty, tin1, tin2, tout;
    float x1, y1, x2, y2;
    List < PointF > line = new List < PointF > ( );
    if ( vertices. Length < = 1) /* nothing to clip */
        return vertices;
```

遍历多边形点的集合，根据后一个点 X 坐标值 x2，与前一个点 X 坐标值 x1 的差值 deltax，与 0 进行比较，若值为 0，且前一个点 X 坐标值大于 0，设置 deltax 为负的零值，若前一个点 X 坐标值小于 0，设置 deltax 为正的零值；并用同样的方式判断其 Y 坐标值，如下所示：

```
for ( int i = 0; i < vertices. Length - 1; i + + )
{
    x1 = vertices[ i]. X;
    y1 = vertices[ i]. Y;
    x2 = vertices[ i + 1]. X;
    y2 = vertices[ i + 1]. Y;
    deltax = x2 - x1;
    if ( deltax = = 0) { deltax = ( x1 > 0) ? - NearZero : NearZero; }
    deltay = y2 - y1;
    if ( deltay = = 0) { deltay = ( y1 > 0) ? - NearZero : NearZero;}
```

若 deltax 大于 0，设置点在 X 坐标方向为从左到右，即 xin 为 0，xout 为图像宽度 width，否则设置点的方向为从右往左，即 xin 为图像的宽度 width，xout 为 0；用同样的方式判断 Y 坐标方向：

```
if ( deltax > 0) {xin = 0; xout = width; }
else{ xin = width; xout = 0; }
if ( deltay > 0) {yin = 0; yout = height; }
else{yin = height; yout = 0; }
```

计算 x1、y1 的 tinx 值、tiny 值，判断点是在 X 方向上，还是在 Y 方向上。若在 X 方向上先接触，设置 tin1 值为 tinx、tin2 为 tiny；若在 Y 方向上先接触，设置 tin1 为 tiny、tin2 为 tinx：

```
tinx = ( xin - x1)/deltax;
tiny = ( yin - y1)/deltay;
if ( tinx < tiny) {tin1 = tinx; tin2 = tiny; }
```

else{ tin1 = tiny; tin2 = tinx; }

若 tin1 不大于 1，且 tin1 大于 0，直接将以 xin、yin 为坐标的点加入 line 中；若 tin2 不大于 1 则再计算在 out 方向上的比值 tout：

```
if (1 > = tin1)
{
    if (0 < tin1)
        line. Add( new PointF( xin, yin) );
    if (1 > = tin2)
    {
        toutx = (xout - x1)/deltax;
        touty = (yout - y1)/deltay;
        tout = (toutx < touty) ? toutx : touty;
```

若 tout 大于 0 或者 tin2 大于 0，若 tin2 不大于 tout，且 tin2 大于 0，tinx 大于 tiny，新建一个以 xin 做 X 坐标值，y1 + tinx * deltay 为 Y 坐标值的点，加入到 line 集合中，其余条件下判断方式类似：

```
if (0 < tin2 || 0 < tout)
{
    if (tin2 < = tout)
    {
        if (0 < tin2)
        {
            if (tinx > tiny)
                line. Add( new PointF( xin, y1 + tinx * deltay) );
            else
                line. Add( new PointF( x1 + tiny * deltax, yin) );
        }
        if (1 > tout)
        {
            if (toutx < touty)
                line. Add( new PointF( xout, y1 + toutx * deltay) );
            else
                line. Add( new PointF( x1 + touty * deltax, yout) );
        }
        else
            line. Add( new PointF( x2, y2) );
    }
    else
    {
        if (tinx > tiny)
            line. Add( new PointF( xin, yout) );
        else
```

```
            line.Add(new PointF(xout, yin));
          }
        }
      }
    }
    if (line.Count > 0)
        line.Add(new PointF(line[0].X, line[0].Y));
    return line.ToArray();
}
```

5.6 主　题

在制图表达中，往往需要根据要素的不同类型采用不同的表达方式。例如一个道路要素集，有高速公路、一级公路、乡间小路、铁路等几种不同类型，需要采用不同的格式（Style）绘制。SharpMap.Rendering.Thematics 名称空间中的对象就是用于这一目的，即通过主题（Theme）对象提供的功能，可以根据要素对象（FeatureDataRow）返回此要素的样式。

SharpMap 实现了 GradientTheme 和 CustomTheme 主题类。GradientTheme 提供一种可渐变颜色的渲染方式，而 CustomTheme 通过委托的方式，实现可配置的绘制风格。这两个类都直接或间接地实现了 ITheme 接口。

5.6.1　ITheme 接口

ITheme 接口只有一个函数，就是根据输入的要素对象，返回相应的绘制格式对象，代码如下：

```
IStyle GetStyle(FeatureDataRow attribute);
```

5.6.2　色彩混合类 ColorBlend

ColorBlend 类定义了一个颜色数组（Color []）及位置数组（float []），用于多色渐变时的色彩解析。该类的功能是将颜色与数值对应起来，通过调用 GetColor 函数，系统根据函数的输入值（float 类型）及 ColorBlend 对象内部颜色数组、位置数组成员的相应值，经过线性插值生成一个中间颜色并返回。

5.6.2.1　ColorBlend 的数据成员

A　_maximum

它是渐变范围的最大值，代码如下：

```
private float _maximum = float.NaN;
```

B　_minimum

它是渐变范围的最小值，代码如下：

```
private float _minimum = float. NaN;
```

5.6.2.2　ColorBlend 类的属性

A　Colors

获取设置存储渐变色彩的数组,代码如下:

```
private Color[ ] _Colors;
public Color[ ] Colors
{
    get { return _Colors; }
    set { _Colors = value; }
}
```

B　Positions

获取和设置存储渐变点位置的数组,数组中的元素表示在颜色渐变时,每个渐变梯度所占的百分比,故此数组中元素的值在 0~1 之间,代码如下:

```
private float[ ] _Positions;
public float[ ] Positions
{
    get { return _Positions; }
    set
    { _Positions = value;
      if ( value = = null ) _minimum = _maximum = float. NaN;
      else { _minimum = value[0]; _maximum = value[value. GetUpperBound(0)]; }
    }
}
```

C　Rainbow7

获取一个预定义的七色渐变色彩混合对象,色彩从红色渐变到紫色,色彩渐变的间隔为 1/6。这是一个只读属性,首先实例化一个色彩混合对象 cb,设置其_ Positions 值为新建的 float 类型的数组,然后设置数组元素的值为 1/6 的倍数;最后新建一个 Color 类型的数组,元素的值分别为红、橙、黄、绿、青、蓝、紫这些色彩结构,并赋予 cb. Colors 属性,返回 cb 对象,代码如下:

```
public static ColorBlend Rainbow7
{
get
{
    ColorBlend cb = new ColorBlend( );
    cb. _Positions = new float[7];
    for ( int i = 1; i < 7; i + + )cb. Positions[i] = i/6f;
    cb. Colors = new[ ]{Color. Red,Color. Orange,Color. Yellow,Color. Green,
    Color. Blue,Color. Indigo,Color. Violet};
```

5.6 主题

```
        return cb;
    }
}
```

D Rainbow5

获取一个预定义的五色渐变的色彩混合对象，色彩从红色渐变到蓝色，色彩渐变间隔为 1/4。这里新建一个色彩数组，值分别为红、黄、绿、青、蓝，再新建一个位置数组，值为 1/4 的倍数，并将这两个数组作为参数，新建一个色彩混合对象并返回，代码如下：

```csharp
public static ColorBlend Rainbow5
{
    get
    {
        return new ColorBlend ( new [ ] {Color.Red, Color.Yellow, Color.Green, Color.Cyan,
            Color.Blue},
            new[] {0f, 0.25f, 0.5f, 0.75f, 1f} );
    }
}
```

E BlackToWhite

获取预定义的从黑色到白色的二色渐变的色彩混合对象。首先新建一个色彩数组，值为黑色和白色，接着新建一个位置数组，值为 0 和 1，将这两个数组作为参数新建色彩混合对象并返回，代码如下：

```csharp
public static ColorBlend BlackToWhite
{
    get { return new ColorBlend(new[] {Color.Black, Color.White}, new[] {0f, 1f} ); }
}
```

F WhiteToBlack

获取一个预定义的从白色到黑色的二色渐变的色彩混合对象。生成方式与 BlackToWhite 相似，代码如下：

```csharp
public static ColorBlend WhiteToBlack
{
    get { return new ColorBlend(new[] {Color.White, Color.Black}, new[] {0f, 1f} ); }
}
```

G RedToGreen

获取一个预定义的从红色到绿色的二色渐变的色彩混合对象。生成方式与 BlackToWhite 相似，代码如下：

```csharp
public static ColorBlend RedToGreen
{
    get { return new ColorBlend(new[] {Color.Red, Color.Green}, new[] {0f, 1f} ); }
}
```

H GreenToRed

获取一个预定义的从绿色到红色的二色渐变的色彩混合对象。生成方式与 BlackToWhite 相似，代码如下：

```
public static ColorBlend GreenToRed
{
    get { return new ColorBlend(new[] {Color.Green, Color.Red}, new[] {0f, 1f}); }
}
```

I BlueToGreen

获取一个预定义的从蓝色到绿色的二色渐变的色彩混合对象。生成方式与 BlackToWhite 相似，代码如下：

```
public static ColorBlend BlueToGreen
{
    get { return new ColorBlend(new[] {Color.Blue, Color.Green}, new[] {0f, 1f}); }
}
```

J GreenToBlue

获取一个从绿色到蓝色的二色渐变的色彩混合对象。生成方式与 BlackToWhite 相似，代码如下：

```
public static ColorBlend GreenToBlue
{
    get { return new ColorBlend(new[] {Color.Green, Color.Blue}, new[] {0f, 1f}); }
}
```

K RedToBlue

获取一个预定义的从红色到蓝色的二色渐变的色彩混合对象。生成方式与 BlackToWhite 相似，代码如下：

```
public static ColorBlend RedToBlue
{
    get { return new ColorBlend(new[] {Color.Red, Color.Blue}, new[] {0f, 1f}); }
}
```

L BlueToRed

获取一个预定义的从蓝色到红色的二色渐变的色彩混合对象。生成方式与 BlackToWhite 相似，代码如下：

```
public static ColorBlend BlueToRed
{
    get { return new ColorBlend(new[] {Color.Blue, Color.Red}, new[] {0f, 1f}); }
}
```

5.6.2.3 ColorBlend 类的方法

A ColorBlend ( )

默认构造函数。这里什么也没做，代码如下：

```
internal ColorBlend(){}
```

B ColorBlend (Color [ ] colors, float [ ] positions)

构造函数的重载，参数为色彩数组 colors、位置数组 positions，用于在构造函数中设置_Colors 成员和 Positions 属性，以初始化新的色彩混合对象，代码如下：

```
public ColorBlend(Color[] colors, float[] positions)
{
    _Colors = colors;
    Positions = positions;
}
```

C GetColor

此方法获取指定位置的颜色。首先，若色彩渐变的最小值_minimum 为空，则抛出异常，提示未设置色彩位置；若色彩的个数与位置的个数不相等，则抛出异常，提示两者必须相等；若色彩的个数小于 2，则抛出异常，提示色彩混合对象至少需要两种颜色，代码如下：

```
public Color GetColor(float pos)
{
    if (float.IsNaN(_minimum))
        throw (new ArgumentException("Positions not set"));
    if (_Colors.Length != _Positions.Length)
        throw (new ArgumentException("Colors and Positions arrays must be of equal length"));
    if (_Colors.Length < 2)
        throw (new ArgumentException("At least two colors must be defined in the ColorBlend"));
```

新建整形变量 i，初始值为 1，用于保存指定位置的索引号；然后计算指定位置所占的比例 frac，并调用 Math.Max、Math.Min 方法确保值在 0~1 之间，然后分别计算色彩在 R、G、B、A 色彩分量上的值，计算方式为当前色彩所占比例为 frac，前一位置所占比例为 (1 - frac)，最后调用 Color.FromArgb 方法生成新的色彩对象并返回：

```
    int i = 1;
    while (i < _Positions.Length && _Positions[i] < pos)
        i++;
    float frac = (pos - _Positions[i - 1])/(_Positions[i] - _Positions[i - 1]);
    frac = Math.Max(frac, 0.0f);
    frac = Math.Min(frac, 1.0f);
    int R = (int) Math.Round((_Colors[i - 1].R * (1 - frac) + _Colors[i].R * frac));
    int G = (int) Math.Round((_Colors[i - 1].G * (1 - frac) + _Colors[i].G * frac));
```

```
int B = (int) Math.Round((_Colors[i - 1].B * (1 - frac) + _Colors[i].B * frac));
int A = (int) Math.Round((_Colors[i - 1].A * (1 - frac) + _Colors[i].A * frac));
return Color.FromArgb(A, R, G, B);
}
```

D ToBrush

此方法将色彩混合对象转换为线性渐变画刷对象。首先新建一个线性画刷对象 br，参数为画刷的大小 rectangle，从黑色渐变到黑色，渐变角度为 angle；新建一个 .NET 的色彩混合对象 cb，设置 cb.Colors 属性值为_Colors；然后计算每个位置值，为当前位置所占渐变最大值和最小值之差的比值，最后设置 cb.Position 属性为重新计算后的位置数组，设置线性渐变画刷对象 br 的 InterpolationColors 属性值为 cb，返回此画刷，代码如下：

```
public LinearGradientBrush ToBrush(Rectangle rectangle, float angle)
{
    LinearGradientBrush br = new LinearGradientBrush(rectangle, Color.Black, Color.Black, angle, true);
    System.Drawing.Drawing2D.ColorBlend cb = new System.Drawing.Drawing2D.ColorBlend();
    cb.Colors = _Colors;
    float[] positions = new float[_Positions.Length];
    float range = _maximum - _minimum;
    for (int i = 0; i < _Positions.Length; i++) positions[i] = (_Positions[i] - _minimum) / range;
    cb.Positions = positions;
    br.InterpolationColors = cb;
    return br;
}
```

E TwoColors

此方法返回二色渐变的色彩混合对象，参数为渐变起始色 fromColor、渐变终止色 toColor。这里直接将这两个色彩作为初始值，新建一个色彩数组，接着新建一个位置数组，值为 0、1，将这两个数组作为参数，新建色彩混合对象并返回，代码如下：

```
public static ColorBlend TwoColors(Color fromColor, Color toColor)
{
    return new ColorBlend(new[] {fromColor, toColor}, new[] {0f, 1f});
}
```

F ThreeColors

此方法返回三色渐变的色彩混合对象，参数为渐变起始色 fromColor、中间色 middleColor、渐变终止色 toColor，同样的这里将三个参数作为初始值，新建一个颜色数组，接着新建一个位置数组，值为 0、0.5、1，将这两个数组作为参数新建色彩混合对象并返回，代码如下：

```
public static ColorBlend ThreeColors(Color fromColor, Color middleColor, Color toColor)
```

```
        return new ColorBlend(new[]{fromColor, middleColor, toColor}, new[]{0f, 0.5f, 1f});
    }
```

5.6.3 GradientThemeBase 类

这是一个抽象类，封装了渐变色渲染的功能。其工作方式是根据空间要素的某个数值型属性的值构造相应的 Style。后面的 GradientTheme 是该类的派生类。

5.6.3.1 GradientThemeBase 类的属性

A　Min

设置和获取渐变的最小值，代码如下：

```
private double _min;
public double Min
{
    get { return _min; }
    set { _min = value; }
}
```

B　Max

设置和获取渐变的最大值，代码如下：

```
private double _max;
public double Max
{
    get { return _max; }
    set { _max = value; }
}
```

C　MinStyle

当空间要素相关属性值小于等于最小值（Min）时所使用的符号，代码如下：

```
private IStyle _minStyle;
public IStyle MinStyle
{
    get { return _minStyle; }
    set { _minStyle = value; }
}
```

D　MaxStyle

当空间要素相关属性值大于等于最大值（Max）时所使用的符号，代码如下：

```
private IStyle _maxStyle;
public IStyle MaxStyle
{
```

```
        get { return _maxStyle; }
        set { _maxStyle = value; }
    }
```

E   TextColorBlend

设置和获取绘制文本所使用的混合色,代码如下:

```
    private ColorBlend _TextColorBlend;
    public ColorBlend TextColorBlend
    {
        get { return _TextColorBlend; }
        set { _TextColorBlend = value; }
    }
```

F   LineColorBlend

设置和获取绘制线操作时所使用的混合色,代码如下:

```
    private ColorBlend _LineColorBlend;
    public ColorBlend LineColorBlend
    {
        get { return _LineColorBlend; }
        set { _LineColorBlend = value; }
    }
```

G   FillColorBlend

设置和获取填充操作时所使用的混合色,代码如下:

```
    private ColorBlend _FillColorBlend;
    public ColorBlend FillColorBlend
    {
        get { return _FillColorBlend; }
        set { _FillColorBlend = value; }
    }
```

5.6.3.2 GradientThemeBase 类的方法

A   GradientThemeBase

构造函数,需传入的参数分别是渐变最小值 minValue、渐变最大值 maxValue、最小样式 minStyle、最大样式 maxStyle,在构造函数中分别存储于对应的成员变量中,代码如下:

```
    protected GradientThemeBase(double minValue, double maxValue, IStyle minStyle, IStyle maxStyle)
    {
        _min = minValue;
        _max = maxValue;
        _maxStyle = maxStyle;
```

```
        _minStyle = minStyle;
    }
```

B　CalculateVectorStyle

此方法计算矢量样式，根据输入参数 value 的值，计算出与类数据成员 _min 及 _max 之间的比例，然后按比例构造一个介于最小（min）与最大（max）之间的格式对象，代码如下：

```
protected VectorStyle CalculateVectorStyle(VectorStyle min, VectorStyle max, double value)
{
    VectorStyle style = new VectorStyle();
```

调用 Fraction 计算 value 值对应的比值：

```
double dFrac = Fraction(value);
float fFrac = Convert.ToSingle(dFrac);
```

根据 dFrac 的值是否大于 0.5 确定 style 是否可用，若大于 0.5，可用性与最小样式一致，反之，与最大样式一致：

```
style.Enabled = (dFrac > 0.5 ? min.Enabled : max.Enabled);
style.EnableOutline = (dFrac > 0.5 ? min.EnableOutline : max.EnableOutline);
```

若填充色混合不为空，设置样式的填充为单色画刷 SolidBrush 对象，画刷颜色为 _FillColorBlend.GetColor 方法得到的值；若填充色混合为空，则设置样式填充为调用 InterpolateBrush 方法后的画刷对象：

```
if (_FillColorBlend != null)
    style.Fill = new SolidBrush(_FillColorBlend.GetColor(fFrac));
else if (min.Fill != null && max.Fill != null)
    style.Fill = InterpolateBrush(min.Fill, max.Fill, value);
style.Line = InterpolatePen(min.Line, max.Line, value);
if (_LineColorBlend != null)
    style.Line.Color = _LineColorBlend.GetColor(fFrac);
```

若最小样式和最大样式的边线不为空，设置样式的边线为调用 InterpolatePen 方法得到的画笔对象：

```
if (min.Outline != null && max.Outline != null)
    style.Outline = InterpolatePen(min.Outline, max.Outline, value);
```

调用 InterpolateDouble 方法计算样式的最小可视化值和最大可视化值，并设置 style.MinVisible 和 style.MaxVisible：

```
style.MinVisible = InterpolateDouble(min.MinVisible, max.MinVisible, value);
style.MaxVisible = InterpolateDouble(min.MaxVisible, max.MaxVisible, value);
```

根据计算的百分比是否大于 0.5 设置样式的符号，若大于 0.5 为最小样式的符号，若小于 0.5 为最大样式的符号；符号的偏移值也是取决于百分比是否大于 0.5，若大于则为最小样式的偏移，反之为最大样式的偏移：

```
style.Symbol = (dFrac > 0.5 ? min.Symbol : max.Symbol);
style.SymbolOffset = (dFrac > 0.5 ? min.SymbolOffset : max.SymbolOffset);
//We don't interpolate the offset but let it follow the symbol instead
```

最后，调用 InterpolateFloat 计算符号的比例，设置 style.SymbolScale 值，并返回样式对象：

```
style.SymbolScale = InterpolateFloat(min.SymbolScale, max.SymbolScale, value);
return style;
}
```

C CalculateLabelStyle

此方法与 CalculateVectorStyle 类似，根据 value 值，构造注记格式对象，代码如下：

```
protected LabelStyle CalculateLabelStyle(LabelStyle min, LabelStyle max, double value)
{
    LabelStyle style = new LabelStyle();
    style.CollisionDetection = min.CollisionDetection;
```

调用 InterpolateBool 计算样式的可用性并设置 style.Enabled 属性；调用 InterpolateFloat 计算字体大小并设置 float.FontSize 属性；实例化一个新的字体，设置 style.Font 属性：

```
style.Enabled = InterpolateBool(min.Enabled, max.Enabled, value);
float.FontSize = InterpolateFloat(min.Font.Size, max.Font.Size, value);
style.Font = new Font(min.Font.FontFamily, FontSize, min.Font.Style);
```

若最大样式背景色和最小样式背景色不为空，调用 InterpolateBrush 方法计算背景色，设置 style.BackColor 属性；若文本色混合不为空，调用 _LineColorBlend.GetColor 方法获取颜色，设置样式的前景色 style.ForeColor 属性，否则调用 InterpolateColor 方法计算颜色，设置样式的前景色 style.ForeColor 属性；若最大样式的光环和最小样式的光环不为空，调用 InterpolatePen 方法计算画笔，设置 style.Halo 属性：

```
if (min.BackColor != null && max.BackColor != null)
    style.BackColor = InterpolateBrush(min.BackColor, max.BackColor, value);
if (_TextColorBlend != null)
    style.ForeColor =
        _LineColorBlend.GetColor(Convert.ToSingle(Fraction(value)));
else
    style.ForeColor = InterpolateColor(min.ForeColor, max.ForeColor, value);
if (min.Halo != null && max.Halo != null)
    style.Halo = InterpolatePen(min.Halo, max.Halo, value);
```

分别调用 InterpolateDouble、InterpolateDouble、InterpolateFloat 计算最小可见性、最大可见性和偏移值，最后返回样式对象：

```
style.MinVisible = InterpolateDouble(min.MinVisible, max.MinVisible, value);
```

5.6 主题

```
        style.MaxVisible = InterpolateDouble(min.MaxVisible, max.MaxVisible, value);
        style.Offset = new PointF(InterpolateFloat(min.Offset.X, max.Offset.X, value),
                                  InterpolateFloat(min.Offset.Y, max.Offset.Y, value));
        return style;
    }
```

D Fraction

计算输入参数 attr 相对于 _min、_max 的比值，代码如下：

```
protected double Fraction(double attr)
{
    if (attr < _min) return 0;
    if (attr > _max) return 1;
    return (attr - _min) / (_max - _min);
}
```

E InterpolateBool

转换 Bool 类型，参数为 Bool 类型的 min、max 和一个 double 类型的值 attr。首先调用 Fraction 方法计算百分比，若百分比大于 0.5 返回 max，否则返回 min，代码如下：

```
protected bool InterpolateBool(bool min, bool max, double attr)
{
    double frac = Fraction(attr);
    if (frac > 0.5) return max;
    else return min;
}
```

F InterpolateFloat

修正 float 类型值的计算方法，代码如下：

```
protected float InterpolateFloat(float min, float max, double attr)
{
    return Convert.ToSingle((max - min) * Fraction(attr) + min);
}
```

G InterpolateDouble

修正 double 类型值的计算方法，代码如下：

```
protected double InterpolateDouble(double min, double max, double attr)
{
    return (max - min) * Fraction(attr) + min;
}
```

H InterpolateBrush

根据输入数值，构造画刷的计算方法，首先判断传入的参数是否是单色画刷 Solid-Brush 类型，若不是，抛出异常，提示只支持单色画刷类型；否则，新建一个单色画刷对象，画刷的颜色为调用 InterpolateColor 方法计算的结果，代码如下：

```
protected SolidBrush InterpolateBrush(Brush min, Brush max, double attr)
{
    if (min.GetType() != typeof(SolidBrush) || max.GetType() !=
typeof(SolidBrush))
        throw (new ArgumentException("Only SolidBrush brushes are supported in
        GradientTheme"));
    return new SolidBrush(InterpolateColor((min as SolidBrush).Color, (max as
    SolidBrush).Color, attr));
}
```

Ⅰ　InterpolatePen

根据输入数值，构造画笔的计算方法，首先判断传入的参数类型是否为单色画笔，若不是则抛出异常，提示只支持单色画笔；否则实例化一个画笔对象，画笔的颜色和大小分别是 InterpolateColor 和 InterpolateFloat 计算的结果，代码如下：

```
protected Pen InterpolatePen(Pen min, Pen max, double attr)
{
    if (min.PenType != PenType.SolidColor || max.PenType !=
PenType.SolidColor)
        throw (new ArgumentException("Only SolidColor pens are supported in
        GradientTheme"));
    Pen pen = new Pen(InterpolateColor(min.Color, max.Color, attr),
    InterpolateFloat(min.Width, max.Width, attr));
```

调用 Fraction 方法计算参数值的百分比；调用 InterpolateFloat 方法计算要将此钢笔的粗细减为一半的斜接长度的比率限制，设置 pen.MiterLimit 属性；根据百分比的值的大小是否大于0.5，分别设置笔画起点使用的形状的类型 pen.StartCap、笔画末端使用的形状的类型 pen.EndCap、在形状轮廓的顶点处使用的接合类型 pen.LineJoin、生成的虚线的样式 pen.DashStyle 属性：

```
double frac = Fraction(attr);
pen.MiterLimit = InterpolateFloat(min.MiterLimit, max.MiterLimit, attr);
pen.StartCap = (frac > 0.5 ? max.StartCap : min.StartCap);
pen.EndCap = (frac > 0.5 ? max.EndCap : min.EndCap);
pen.LineJoin = (frac > 0.5 ? max.LineJoin : min.LineJoin);
pen.DashStyle = (frac > 0.5 ? max.DashStyle : min.DashStyle);
```

若最大样式和最小样式的虚线样式不为空，根据百分比是否大于0.5确定虚线中交替出现的短划线和空白区域的长度 DashPattern。然后还是根据百分比的值，确定线的起点到短划线图案起始处的距离 DashOffset、虚线的末端类型 DashCap、复合钢笔的值数组 CompoundArray、直线起点使用的自定义线帽 CustomStartCap、直线终点使用的自定义线帽 CustomEndCap，最后返回画笔对象：

```
        if ( min. DashStyle = = DashStyle. Custom && max. DashStyle = = DashStyle. Custom)
            pen. DashPattern = (frac > 0.5 ? max. DashPattern : min. DashPattern);
        pen. DashOffset = (frac > 0.5 ? max. DashOffset : min. DashOffset);
        pen. DashCap = (frac > 0.5 ? max. DashCap : min. DashCap);
        if ( min. CompoundArray. Length > 0 && max. CompoundArray. Length > 0)
            pen. CompoundArray = (frac > 0.5 ? max. CompoundArray : min. CompoundArray);
        pen. Alignment = (frac > 0.5 ? max. Alignment : min. Alignment);
        return pen;
    }
```

J InterpolateColor

根据输入数值，构造颜色的计算方法，传入的参数是最小样式的颜色 minCol、最大样式的颜色 maxCol 和值 attr，这里首先计算值的百分比，若百分比等于 1，返回最大样式的颜色；若百分比等于 0，返回最小样式颜色；若在 0 和 1 之间，分别修正 R、G、B、A 值，调用 Color. FromArgb 方法生成新的色彩对象并返回，代码如下：

```
        protected Color InterpolateColor( Color minCol, Color maxCol, double attr)
        {
            double frac = Fraction( attr);
            if (frac = = 1)
                return maxCol;
            else if (frac = = 0)
                return minCol;
            else
            {
                double r = (maxCol. R - minCol. R) * frac + minCol. R;
                double g = (maxCol. G - minCol. G) * frac + minCol. G;
                double b = (maxCol. B - minCol. B) * frac + minCol. B;
                double a = (maxCol. A - minCol. A) * frac + minCol. A;
                if (r > 255) r = 255;
                if (g > 255) g = 255;
                if (b > 255) b = 255;
                if (a > 255) a = 255;
                return Color. FromArgb((int)a, (int)r, (int)g, (int)b);
            }
        }
```

K GetStyle

根据输入的要素对象，构造出相应的格式对象。

先调用 GetAttributeValue 方法获取数值，再根据该数值构造格式对象，代码如下：

```
        public virtual IStyle GetStyle( FeatureDataRow row)
        {
            double attr = 0;
```

```
        try
        {
            attr = GetAttributeValue(row);
        }
    catch
        {
            throw new ApplicationException(
                "Invalid Attribute type in Gradient Theme - Couldn't parse
                attribute (must be numerical)");
        }
        if (_minStyle.GetType() != _maxStyle.GetType())
            throw new ArgumentException("MinStyle and MaxStyle must be of the same type");
```

然后根据最小样式的类型的名称，计算并返回不同的样式，当样式的名称为"SharpMap.Styles.VectorStyle"时，调用 CalculateVectorStyle 方法计算矢量样式；当样式的名称为"SharpMap.Styles.LabelStyle"时，调用 CalculateLabelStyle 方法计算标注样式；其他情况则抛出异常，提示渐变色主题目前仅支持矢量样式和标注样式：

```
    switch (MinStyle.GetType().FullName)
    {
        case "SharpMap.Styles.VectorStyle":
            return CalculateVectorStyle(MinStyle as VectorStyle, MaxStyle as
                VectorStyle, attr);
        case "SharpMap.Styles.LabelStyle":
            return CalculateLabelStyle(MinStyle as LabelStyle, MaxStyle as
                LabelStyle, attr);
        default:
            throw new ArgumentException(
                "Only SharpMap.Styles.VectorStyle and
            SharpMap.Styles.LabelStyle are supported for the gradient theme");
    }
```

5.6.4 GradientTheme 类

GradientTheme 是渐变主题类，派生自 GradientThemeBase。该类中多定义了一个字符串数据成员，用于记录字段名称，这个字段的值被用于构造 Style 对象，参见 GradientThemeBase 类的 GetStyle 函数，并提供了初始化、访问该成员的构造函数及属性。代码如下：

```
    private string _ColumnName;
```

此外，重载了 GetAttributeValue 函数，代码如下：

```
    protected override double GetAttributeValue(FeatureDataRow row)
    {
```

```
        return Convert.ToDouble(row[_ColumnName]);
    }
```

由于该类很简单,书中不再赘述。

5.6.5 CustomTheme 类

该类可以方便程序员通过代理来构建自定义主题类,而不需要从 ITheme 派生新类。

5.6.5.1 CustomTheme 类的属性

A DefaultStyle

获取和设置默认的样式,代码如下:

```
private IStyle _DefaultStyle;
public IStyle DefaultStyle
{
    get { return _DefaultStyle; }
    set { _DefaultStyle = value; }
}
```

B StyleDelegate

获取和设置决定要素样式的自定义委托,代码如下:

```
public delegate IStyle GetStyleMethod(FeatureDataRow dr);
private GetStyleMethod _getStyleDelegate;
public GetStyleMethod StyleDelegate
{
    get { return _getStyleDelegate; }
    set { _getStyleDelegate = value; }
}
```

5.6.5.2 CustomTheme 类的方法

A CustomTheme

构造方法,传入的参数是用户自定义的委托,代码如下:

```
public CustomTheme(GetStyleMethod getStyleMethod)
{
    _getStyleDelegate = getStyleMethod;
}
```

B GetStyle

获取样式,参数为要素数据行对象 row。这里调用委托获取要素数据行的样式,若样式不为空,返回此样式,若为空,则返回默认的样式,代码如下:

```
public IStyle GetStyle(FeatureDataRow row)
{
```

```
        IStyle style = _getStyleDelegate(row);
        if (style != null)
            return style;
        else
            return _DefaultStyle;
}
```

复习思考题

5-1 理解 SharpMap 中如何实现地图注记的功能,画出算法流程图。
5-2 SharpMap 中如何对矢量图层进行渲染?试画出算法流程图。
5-3 SharpMap 中主题的作用是什么?

第6章 样 式

样式（Style）用于绘制空间数据。有些系统中（如 ArcGIS），称为符号（symbol）。

GIS 中，系统在绘制图形时，都需要使用某种样式来显示一个要素。比如一个要素用来表示公路，一个要素用来表示河流，它们在文件存储时没有太大的区别，而在绘制时，由于使用了不同的样式，就可以具有更好的地图表达效果。

目前 SharpMap 中，实现了两个 Style 类，即 VectorStyle 和 LabelStyle。VectorStyle 是用来绘制矢量要素的，点、线或面都可以使用这个类的对象来进行绘制，但是不同的几何类型使用 VectorStyle 对象中的不同成员。查看 VectorStyle 的几个成员，很容易明白各个成员的意义。LabelStyle 用来绘制标注文本，可以设置其字体、颜色、大小、背景色以及水平对齐和垂直对齐方式等，应该说对一般的文本显示来说够用了。

图 6-1 所示是 SharpMap 的 Styles 类图。

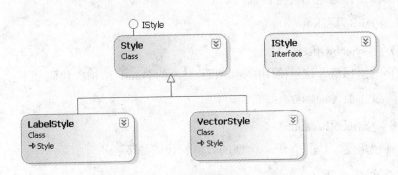

图 6-1 SharpMap 的 Styles 类图

6.1 矢量图层样式 VectorStyle

矢量图层的样式有一些共性，这些共性来自于继承 IStyle 接口的结果，这些共性包括在什么比例尺区间内显示、是否允许使用样式。矢量图形分为点、线、面三种类型，不过在 SharpMap 中并没有为每种几何数据类型划分出单独的样式。

6.1.1 VectorStyle 的数据成员

6.1.1.1 DefaultSymbol

默认的点符号所使用的图像，在构造函数中初始化其值，代码如下：

```
public static readonly Image DefaultSymbol;
```

6.1.1.2　_fillStyle

填充样式，其实就是.NET类库的画刷类Brush，代码如下：

```
private Brush _fillStyle;
```

6.1.1.3　_lineStyle

线的样式，使用的是.NET类库的画笔类pen，代码如下：

```
private Pen _lineStyle;
```

6.1.1.4　_outline

符号的外边框是否显示，这是一个布尔值，代码如下：

```
private bool _outline;
```

6.1.1.5　_outlineStyle

符号外边框的线的样式，这里使用的是.NET类库的画笔pen，代码如下：

```
private Pen _outlineStyle;
```

6.1.1.6　_symbol

用于绘制点数据的符号，这里使用的是.NET类库的栅格图像Image类，所以目前只能使用栅格图像来作为地图符号，代码如下：

```
private Image _symbol;
```

6.1.1.7　_symbolOffset

符号偏移的坐标值，这里使用的是.NET类库的PointF结构，代码如下：

```
private PointF _symbolOffset;
```

6.1.1.8　_symbolRotation

符号的旋转角度，是一个float类型的数值，代码如下：

```
private float _symbolRotation;
```

6.1.1.9　_symbolScale

符号的比例，是一个float类型的数值，代码如下：

```
private float _symbolScale;
```

6.1.2　VectorStyle 的属性

VectorStyle 类中的属性如 EnbleOutLine、Fill、Line、OutLine、Symbol、SymbolOffSet、SymbolRotation、SymbolScale，都是对以上数据成员提供了简单的封装，请读者自行看代码，书中不再赘述。

6.1.3　VectorStyle 的方法

6.1.3.1　静态构造方法

这个构造函数从嵌入的资源里获取栅格类型的图片作为默认符号，代码如下：

```
static VectorStyle()
{
System.IO.Stream rs = Assembly.GetExecutingAssembly().GetManifestResourceStream
("SharpMap.Styles.DefaultSymbol.png");
    if (rs != null) DefaultSymbol = Image.FromStream(rs);
}
```

6.1.3.2 默认构造函数

这个构造函数将符号的外边框线性设置为黑色且默认不显示，符号的填充色也设置为黑色，符号的缩放比例为1，代码如下：

```
public VectorStyle()
{
    Outline = new Pen(Color.Black, 1);
    Line = new Pen(Color.Black, 1);
    Fill = Brushes.Black;
    EnableOutline = false;
    SymbolScale = 1f;
}
```

6.2 标注样式 LabelStyle

标注样式是在绘制地图标注时使用的样式，它决定了标注的外观。

6.2.1 LabelStyle 的数据成员

6.2.1.1 _BackColor

标注的背景颜色，这里使用的是 .NET 的画刷类 Brush 来填充，代码如下：

```
private Brush _BackColor;
```

6.2.1.2 _CollisionBuffer

冲突的缓冲大小，是一个 SizeF 结构的矩形，代码如下：

```
private SizeF _CollisionBuffer;
```

6.2.1.3 _CollisionDetection

是否允许冲突检测，代码如下：

```
private bool _CollisionDetection;
```

6.2.1.4 _Font

标注时所使用的字体，代码如下：

```
private Font _Font;
```

6.2.1.5 _ForeColor

标注时的前景色,代码如下:

```
private Color _ForeColor;
```

6.2.1.6 _Halo

光晕效果,这里使用画笔类 Pen,代码如下:

```
private Pen _Halo;
```

6.2.1.7 _HorisontalAlignment

水平对齐方式,是一个 HorizontalAlignmentEnum 类型的枚举值,代码如下:

```
private HorizontalAlignmentEnum _HorisontalAlignment;
```

6.2.1.8 _Offset

标注的偏移值,代码如下:

```
private PointF _Offset;
```

6.2.1.9 _VerticalAlignment

竖直对齐方式,是一个 VerticalAlignmentEnum 类型枚举值,代码如下:

```
private VerticalAlignmentEnum _VerticalAlignment;
```

6.2.2 LabelStyle 的属性

LabelStyle 的属性提供了对以上数据成员的封装,例如 Font、BackColor、ForeColor;其中 VerticalAlignment、HorizontalAlignment、CollisionBuffer、CollisionDetection、Offset 还使用了.NET 的特性 System.ComponentModel.Category;Halo 除了 System.ComponentModel.Category 特性外,还使用了 System.ComponentModel.Editor 特性,有关特性(Atribute)的知识,请参考.NET 开发帮助。

6.2.3 LabelStyle 的方法

LabelStyle 构造函数中,将绘制字体设置为 Times New Roman,偏移值默认为(0,0),冲突检测设置为 false,冲突缓冲区为 0,水平对齐方式为中央对齐,竖直对齐方式为中间对齐。

```
public LabelStyle()
{
    _Font = new Font("Times New Roman", 12f);
    _Offset = new PointF(0, 0);
    _CollisionDetection = false;
    _CollisionBuffer = new Size(0, 0);
```

　　　　　_ForeColor = Color.Black;
　　　　　_HorisontalAlignment = HorizontalAlignmentEnum.Center;
　　　　　_VerticalAlignment = VerticalAlignmentEnum.Middle;
}

复习思考题

6-1　SharpMap 支持的样式有哪几种？思考矢量图层样式设置的实现方式。

6-2　思考如何对 SharpMap 的样式进行扩展。

第 7 章 数 据

SharpMap 中数据模块定义在 SharpMap.Data 名称空间下，文件位于 SharpMap \ Data 目录下。数据模块主要有以下两个部分：

（1）空间数据库访问对象。位于 SharpMap.Data.Providers 名称空间下，是空间数据的物理层，用于抽象各种不同的空间数据格式，采用的是 Provider 设计模式。包括 IProvider（接口）、ShapeFile、SqlServer2008、MsSql、OleDbPoint、GeometryFeatureProvider、GeometryProvider、DbaseReader 等类。

（2）空间要素模块。位于 SharpMap.Data 名称空间下，是空间数据的逻辑层。各种不同的空间数据，如 shape 格式、PostGIS 格式等，读入内存后就以要素集的形式存在。包括 FeatureDataSet、FeatureDataTable、FeatureTableCollection、FeatureDataRow 等类。

此外，由于空间数据库目前已得到广泛使用，为连接空间数据库，SharpMap 的数据模块还定义了连接类，包括 Connector、ConnectorPool，位于 SharpMap.Data.Providers.Pooling 名称空间下。

Provider 或者 Provider 模式对于很多人应该都不陌生，在各种类库设计中都大量应用了 Provider 模式。目前主流的 GIS 平台的数据模块都是基于 Provider 这样的模式。

Provider 的核心思想在于面向接口编程，也就是说通过接口定义需要的服务，至于服务的实现，则是通过子类继承的方式来实现。就 GIS 数据引擎来说，就是定义对空间数据的操作，例如打开、关闭、读取某个范围内的数据、检索、分析等，然后通过继承这个接口来实现对不同的数据的操作。例如对 Shape 格式的操作和对 PostGIS 格式的操作的具体实现是完全不同的，但其接口一致。这样，系统就能以一种统一的方式（接口）来操纵不同的数据源，这就是实现多源空间数据引擎的方式。

整个 SharpMap.Data.Priviers 名称空间类图如图 7-1 所示，核心为 IProvider 接口，所有的数据源必须实现此接口，例如 ShapeFile、GeometryPrividerder 等。

SharpMap.Data 名称空间类图如图 7-2 所示，其中要素 FeatureDataRow 对象组成要素数据表 FeatureDatatable 对象，要素数据表组成要素数据集对象 FeatureDataSet 对象，这里还定义了要素数据表集合 FeatureTableCollection、要素数据行变化委托类型 FeatureDataRowChangeEventHandler、要素数据行变化事件参数 featureDataRowChangeEventArgs。

7.1 空间数据库连接池技术

数据连接池在 SharpMap.Data.Providers.Pooling 名称空间下，共有两个类 ConnectorPool 和 Connector，Connector 提供了对数据连接的封装，而 ConnectorPool 管理 Connector，控制访问池、共享数据连接者对象。

7.1 空间数据库连接池技术

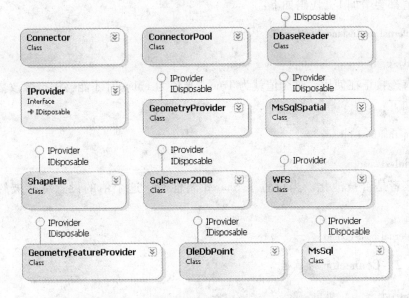

图 7-1　SharpMap.Data.Prividers 名称空间类图

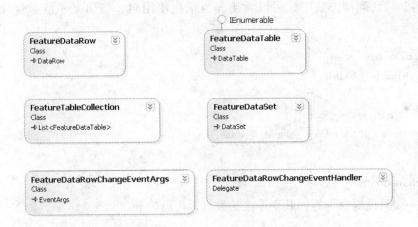

图 7-2　SharpMap.Data 名称空间类图

7.1.1　数据连接对象 Connector

7.1.1.1　Connector 的数据成员

A　InstanceCounter

Connector 类的实例计数器，这个数据成员只是为了调试时使用，并无太大的意义，代码如下：

    private static int InstanceCounter;

B　InstanceNumber

表示当前连接者的实例数量，它实际上使用 InstanceCounter 的计数结果，每实例化一

个对象,值都会增加1,代码如下:

```
internal int InstanceNumber;
```

C InUse

如果有连接正在使用,这个值就为 True,当值为 False 时才能够进行修改数据源等操作,代码如下:

```
internal bool InUse;
```

D Pooled

此连接者是否被池化。当设置了 Shared 属性时,这个值的设置将会被忽略,代码如下:

```
internal bool Pooled;
```

7.1.1.2 Connector 的属性

A Shared

这个属性决定着连接是否能够被共享,该值仅仅能够被 ConnectorPool 的 Request 方法设置。在其 set 访问器方法中,可以看出,当正在使用时,该值是不能被修改的,代码如下:

```
private bool _Shared;
internal bool Shared
{
    get { return _Shared; }
    set { if (! InUse) _Shared = value; }
}
```

B ShareCount

这是一个只读属性,返回共享连接的数目,代码如下:

```
internal int _ShareCount;
internal int ShareCount
{
    get { return _ShareCount; }
}
```

C Provider

获取和设置连接者的数据源,代码如下:

```
private IProvider _Provider;
internal IProvider Provider
{
    get { return _Provider; }
    set
```

```
            }
            if (InUse)
            {
                throw new ApplicationException("Provider cannot be modified if connection is open.");
            }
            _Provider = value;
        }
    }
```

7.1.1.3 Connector 的方法

A Open()

打开数据源。这里调用 IProvider 接口的 Open 方法打开数据源，代码如下：

```
    internal void Open()
    {
        Provider.Open();
    }
```

B Release()

释放连接者。这里首先检查该连接是否被共享，如果共享状态_Shared 为真，检查客户端连接数目是否为 0，如果为 0，就将此连接对象从连接池 SharedConnectors 中移除，并将其加入连接池 PooledConnectors 中，设置 Pooled 值为 true，InUse 值为 False，以供下次有连接请求时使用；若不被共享，就检查此连接的 Pooled 状态，如为真，就将其加入池化连接池 PooledConnectors，否则释放此连接者，调用数据源的 Close() 和 Dispose() 方法完成资源清理和垃圾回收工作，代码如下：

```
    internal void Release()
    {
        if (_Shared)
        {
            if (--_ShareCount == 0)
            {
                ConnectorPool.ConnectorPoolManager.SharedConnectors.Remove(this);
                ConnectorPool.ConnectorPoolManager.PooledConnectors.Add(this);
                Pooled = true;
                InUse = false;
            }
        }
        else // it is a nonshared connector
        {
            if (Pooled)
            {
                InUse = false;
```

```
                ConnectorPool. ConnectorPoolManager. PooledConnectors. Add(this);
            }
            else
            {

                Provider. Close();
                Provider. Dispose();
            }
        }
    }
```

7.1.2 连接池管理 ConnectorPool

7.1.2.1 ConnectorPool 的数据成员

A ConnectorPoolManager

连接池管理者的对象，用于管理 Connector 对象，采用了单件的设计模式（Singleton Pattern），保证了只会有一个连接池管理者实例，代码如下：

```
internal static ConnectorPool ConnectorPoolManager = new ConnectorPool();
```

B PooledConnectors

存放未使用的连接对象的容器，这是一个 List < Connector > 类型的集合，管理着没有使用的"备用"连接，下次有请求时就会返回里面可用的连接者，而避免了到使用时才实例化，从而提高系统效率，代码如下：

```
internal List < Connector > PooledConnectors;
```

C SharedConnectors

共享连接池，这也是一个 List < Connector > 类型的集合，它管理着正在被使用的连接者，代码如下：

```
internal List < Connector > SharedConnectors;
```

7.1.2.2 ConnectorPool 的方法

A ConnectorPool ()

连接池管理者的默认构造函数，注意它的修饰符是 internal，这就意味着仅在当前包中可见。当前包之外的代码不能访问此构造函数。这个构造函数只是初始化了池化连接池和共享连接池，代码如下：

```
internal ConnectorPool()
{
    PooledConnectors = new List < Connector > ();
    SharedConnectors = new List < Connector > ();
}
```

B RequestConnector（IProvider provider，bool Shared）

请求得到一个池化的连接者，这个方法将会对连接池中的连接者进行匹配，匹配成功则返回此连接者，否则就创建一个新的连接者。首先遍历共享连接池 SharedConnectors，将其 ID 编号和参数的 ID 编号进行比较，若相等，则使此连接者的_ShareCount 自加 1，并返回此连接者，代码如下：

```
internal Connector RequestConnector( IProvider provider, bool Shared)
{
    if (Shared)
    {
        foreach (Connector Connector in SharedConnectors)
        {
            if (Connector. Provider. ConnectionID = = provider. ConnectionID)
            {
                Connector._ShareCount + + ;
                return Connector;
            }
        }
    }
}
```

若在共享连接池未匹配到，则遍历空闲连接池 PooledConnectors，若其 ID 编号和参数的 ID 编号相等，将此连接者从连接池中移除；若连接者的 Shared 属性值与参数 Shared 相等，将此连接者加入共享连接池，并设置此连接者的_ShareCount 值为 1，最后设置 InUse 属性为 true，并返回此连接者：

```
foreach (Connector Connector in PooledConnectors)
{
    if (Connector. Provider. ConnectionID = = provider. ConnectionID)
    {
        PooledConnectors. Remove( Connector);
        if (Connector. Shared = Shared)
        {
            SharedConnectors. Add( Connector);
            Connector._ShareCount = 1;
        }
        Connector. InUse = true;
        return Connector;
    }
}
```

若在连接池未匹配到，则实例化一个新的连接者对象 NewConnector，若 Shared 为真，将 NewConnector 加入共享连接池中，设置_ShareCount 为 1，设置 InUse 属性值为 true，并且调用其 Open 方法打开数据源后，返回此连接者：

```
        Connector NewConnector = new Connector(provider, Shared);

    if (Shared)
    {
        SharedConnectors.Add(NewConnector);
        NewConnector._ShareCount = 1;
    }
    NewConnector.InUse = true;
    NewConnector.Open();
    return NewConnector;
}
```

7.2 数据提供接口 IProvider

IProvider 接口是数据源契约，它定义了数据源需提供的功能，接口中的所成员在子类中必须实现。IProvider 中定义了所有数据源操作必需的功能。

7.2.1 IProvider 的属性

7.2.1.1 ConnectionID

数据源的 ID 编号，这是一个唯一值，如文件的文件名或者连接字符串，当使用连接池的时候非常有帮助，如若数据连接池不支持此种数据源，那么 ID 编号为空，代码如下：

```
string ConnectionID { get; }
```

7.2.1.2 IsOpen

返回数据源当前状态是否打开，代码如下：

```
bool IsOpen { get; }
```

7.2.1.3 SRID

空间参考的 ID 编号。这里指的是具体数据源的空间参考的编码，参见附录条目 A，代码如下：

```
int SRID { get; set; }
```

7.2.2 IProvider 的方法

7.2.2.1 GetGeometriesInView

获取指定外包框内的要素的几何图形集合，代码如下：

```
Collection<Geometry> GetGeometriesInView(BoundingBox bbox);
```

7.2.2.2 GetObjectIDsInView

获取与指定外包框相交的对象 ID 的集合，通常这个方法会比 QueryFeatures 方法快很

多,因为相交查询仅仅只是对象的外包框的相交,而且会使用空间索引来加速查询,代码如下:

```
Collection < uint > GetObjectIDsInView( BoundingBox bbox );
```

7.2.2.3 GetGeometryByID
通过几何对象的 ID 编号得到此几何对象,代码如下:

```
Geometry GetGeometryByID( uint oid );
```

7.2.2.4 ExecuteIntersectionQuery
执行相交查询,这个方法会获取与给定几何对象相交的空间要素数据集,代码如下:

```
void ExecuteIntersectionQuery( Geometry geom, FeatureDataSet ds );
```

7.2.2.5 ExecuteIntersectionQuery
执行相交查询,这是一个重载方法,返回和指定外包框相交的空间要素数据集,代码如下:

```
void ExecuteIntersectionQuery( BoundingBox box, FeatureDataSet ds );
```

7.2.2.6 GetFeatureCount
返回数据集里面的要素数目,是一个 Int 类型的整数值,代码如下:

```
int GetFeatureCount( );
```

7.2.2.7 GetFeature
通过要素的编号得到此要素,返回的这个要素是经过包装的 FeatureDataRow 类型的对象(FeatureDataRow 是仿照关系型数据库,类似于数据行的概念,详细信息可以参见本章后面几节内容),代码如下:

```
FeatureDataRow GetFeature( uint RowID );
```

7.2.2.8 GetExtents
得到数据集的外包框矩形,此外包框矩形是数据集中所有要素外包框的并集,代码如下:

```
BoundingBox GetExtents( );
```

7.2.2.9 Open
打开数据源方法,代码如下:

```
void Open( );
```

7.2.2.10 Close
关闭数据源方法,代码如下:

```
void Close( );
```

7.3 DbaseReader 类

DbaseReader 是用于读取 ShapeFile 属性数据的类。属性文件（.dbf）用于记录属性信息。它是一个标准的 DBF 文件，由头文件和实体信息两部分构成。其中文件头部分的长度是不定长的，它主要对 DBF 文件作了一些总体说明；属性文件的实体信息部分就是属性记录，每条记录都是由若干个记录项构成的，因此只要依次循环读取每条记录就可以了。

7.3.1 DbaseReader 类的数据成员

7.3.1.1 _lastUpdate
存储最后一次更新数据的时间，代码如下：

```
private DateTime _lastUpdate;
```

7.3.1.2 _NumberOfRecords
文件中记录的条数，从文件头中读取，代码如下：

```
private int _NumberOfRecords;
```

7.3.1.3 _HeaderLength
文件头的字节数，从文件头中去读，代码如下：

```
private Int16 _HeaderLength;
```

7.3.1.4 _RecordLength
一条记录的字节长度，代码如下：

```
private Int16 _RecordLength;
```

7.3.1.5 _filename
文件的名称，代码如下：

```
private string _filename;
```

7.3.1.6 DbaseColumns
文件的属性的集合，代码如下：

```
private DbaseField[ ] DbaseColumns;
```

7.3.1.7 fs
读取信息的文件流，代码如下：

```
private FileStream fs;
```

7.3.1.8 br
读取信息的二进制文件流，代码如下：

```
private BinaryReader br;
```

7.3.1.9 HeaderIsParsed
文件头是否已经被转换,代码如下:

```
private bool HeaderIsParsed;
```

7.3.1.10 baseTable
文件所属的要素数据表,代码如下:

```
private FeatureDataTable baseTable;
```

7.3.1.11 _FileEncoding
文件的编码方式,代码如下:

```
private Encoding _FileEncoding;
```

7.3.2 DbaseReader 类的属性

7.3.2.1 Encoding
获取和设置转换字符串时的编码方式,代码如下:

```
private Encoding _Encoding;
public Encoding Encoding
{
    get { return _Encoding; }
    set { _Encoding = value; }
}
```

7.3.2.2 IsOpen
文件是否被打开,代码如下:

```
private bool _isOpen;
public bool IsOpen
{
    get { return _isOpen; }
    set { _isOpen = value; }
}
```

7.3.2.3 LastUpdate
最近更新的日期,代码如下:

```
private DateTime _lastUpdate;
public DateTime LastUpdate
{
    get { return _lastUpdate; }
}
```

7.3.2.4 NewTable

创建一个新的数据表,代码如下:

```
internal FeatureDataTable NewTable
{
    get { return baseTable.Clone(); }
}
```

7.3.3 DbaseReader 类的方法

7.3.3.1 DbaseReader(string filename)

构造函数,这里首先检查属性表文件是否存在,若不存在则抛出异常,之后设置文件名称,并设置 HeaderIsParsed 为 false,表示文件头还未被读取,代码如下:

```
public DbaseReader(string filename)
{
    if(!File.Exists(filename))
        throw new FileNotFoundException(String.Format("Could not find file \"{0}\"", filename));
    _filename = filename;
    HeaderIsParsed = false;
}
```

7.3.3.2 Open()

打开属性表文件,初始化读取文件的二进制流对象,若文件头没有转换则调用 ParseDbfHeader() 方法读取文件头,代码如下:

```
public void Open()
{
    fs = new FileStream(_filename, FileMode.Open, FileAccess.Read);
    br = new BinaryReader(fs);
    _isOpen = true;
    if(!HeaderIsParsed) //Don't read the header if it's already parsed
        ParseDbfHeader(_filename);
}
```

7.3.3.3 Close()

关闭函数。这里需要关闭文件读取时用到的流对象并且设置_isOpen 的值为 False,代码如下:

```
public void Close()
{
    br.Close();
    fs.Close();
    _isOpen = false;
}
```

7.3.3.4　Dispose（ ）

实现资源的回收和垃圾清理。这里调用 Close（ ）方法关闭流对象，并将其引用设置为空，代码如下：

```
public void Dispose( )
{
    if ( _isOpen)
        Close( );
    br = null;
    fs = null;
}
```

7.3.3.5　CreateBaseTable（ ）

创建基础数据表，这个表和属性文件存储的数据一样。这里只是创建了一个属性架构一样的要素数据表对象，代码如下：

```
private void CreateBaseTable( )
{
    baseTable = new FeatureDataTable( );
    foreach ( DbaseField dbf in DbaseColumns)
        baseTable.Columns.Add( dbf.ColumnName, dbf.DataType);
}
```

7.3.3.6　CreateDbfIndex < T >（int ColumnId）

将数据表的一个列建立起二叉树索引。可以看到这里只能处理 10000 个以下的属性数据，为之建立起二叉树索引（10000 以后的数据则忽略，并不为其建立索引），函数返回一个 BinaryTree 对象，BinaryTree 类位于 SharpMap.Utilities.Indexing 名称空间下，该类提供了索引功能，用于提高检索效率。由于篇幅关系，本书未对 BinaryTree 类进行讲解，请读者自行研究，代码如下：

```
public BinaryTree < T, UInt32 > CreateDbfIndex < T >(int ColumnId) where T : IComparable < T >
{
    BinaryTree < T, UInt32 > tree = new BinaryTree < T, uint >( );
    for ( uint i = 0; i < ( ( ( _NumberOfRecords > 10000) ? 10000 : _NumberOfRecords); i + + )
        tree.Add( new BinaryTree < T, uint >.ItemValue( (T) GetValue(i, ColumnId), i));
    return tree;
}
```

7.3.3.7　GetDbaseLanguageDriver（byte dbasecode）

根据属性表文件的编码值（代码）转换成对应的编码方式，代码如下：

```
private Encoding GetDbaseLanguageDriver( byte dbasecode)
{
    switch ( dbasecode)
    {
```

```
            case 0x01 :
                return Encoding. GetEncoding(437); //DOS USA code page 437
            case 0x02 :
                return Encoding. GetEncoding(850); // DOS Multilingual code page 850
            case 0x03 :
                return Encoding. GetEncoding(1252); // Windows ANSI code page 1252
```

……以下代码从略,请参见源代码文件
```
        }
    }
```

7.3.3.8 GetFeature(uint oid, FeatureDataTable table)

通过要素的 ID 编号得到这个要素行。首先根据这个 ID 编号的数值得到要素的存储地址,调用 ReadDbfValue() 方法读取属性值并生产新的要素行返回,代码如下:

```
internal FeatureDataRow GetFeature(uint oid, FeatureDataTable table)
{
    if (! _isOpen)
        throw (new ApplicationException("An attempt was made to read from a closed DBF file"));
    if (oid > = _NumberOfRecords)
        throw (new ArgumentException("Invalid DataRow requested at index " + oid. ToString()));
    fs. Seek(_HeaderLength + oid * _RecordLength, 0);
    FeatureDataRow dr = table. NewRow();
    if (br. ReadChar() = = ' * ') //is record marked deleted?
        return null;
    for (int i = 0; i < DbaseColumns. Length; i + +)
    {
        DbaseField dbf = DbaseColumns[i];
        dr[dbf. ColumnName] = ReadDbfValue(dbf);
    }
    return dr;
}
```

7.3.3.9 GetSchemaTable()

得到一个描述属性文件数据列元数据信息的空表,代码如下:

```
public DataTable GetSchemaTable()
{
    DataTable tab = new DataTable();
    // all of common, non "base - table" fields implemented
    tab. Columns. Add("ColumnName", typeof (String));
    tab. Columns. Add("ColumnSize", typeof (Int32));
    tab. Columns. Add("ColumnOrdinal", typeof (Int32));
    tab. Columns. Add("NumericPrecision", typeof (Int16));
```

```
            tab.Columns.Add("NumericScale", typeof(Int16));
            tab.Columns.Add("DataType", typeof(Type));
            tab.Columns.Add("AllowDBNull", typeof(bool));
            tab.Columns.Add("IsReadOnly", typeof(bool));
            tab.Columns.Add("IsUnique", typeof(bool));
            tab.Columns.Add("IsRowVersion", typeof(bool));
            tab.Columns.Add("IsKey", typeof(bool));
            tab.Columns.Add("IsAutoIncrement", typeof(bool));
            tab.Columns.Add("IsLong", typeof(bool));
            foreach(DbaseField dbf in DbaseColumns)
                tab.Columns.Add(dbf.ColumnName, dbf.DataType);
            for(int i = 0; i < DbaseColumns.Length; i++)
            {
                DataRow r = tab.NewRow();
                r["ColumnName"] = DbaseColumns[i].ColumnName;
                r["ColumnSize"] = DbaseColumns[i].Length;
                r["ColumnOrdinal"] = i;
                r["NumericPrecision"] = DbaseColumns[i].Decimals;
                r["NumericScale"] = 0;
                r["DataType"] = DbaseColumns[i].DataType;
                r["AllowDBNull"] = true;
                r["IsReadOnly"] = true;
                r["IsUnique"] = false;
                r["IsRowVersion"] = false;
                r["IsKey"] = false;
                r["IsAutoIncrement"] = false;
                r["IsLong"] = false;
                tab.Rows.Add(r);
            }
            return tab;
        }
```

7.3.3.10 GetValue(uint oid, int colid)

通过要素的 ID 编号和属性的 ID 编号得到属性值。若_isOpen 为假，则抛出异常，提示读取.dbf 文件出错；若 oid 编号大于要素的个数，则抛出异常，提示索引号溢出；若属性列 ID 编号超出合法范围，则抛出异常，提示属性列编号超出范围。然后定位到要素的地址，调用 ReadDbfValue 方法读取属性值，代码如下：

```
        internal object GetValue(uint oid, int colid)
        {
            if(!_isOpen)
                throw(new ApplicationException("An attempt was made to read from a closed DBF file"));
            if(oid >= _NumberOfRecords)
```

```
            throw (new ArgumentException("Invalid DataRow requested at index " + oid.ToString()));
        if (colid >= DbaseColumns.Length || colid < 0)
            throw ((new ArgumentException("Column index out of range")));
        fs.Seek(_HeaderLength + oid * _RecordLength, 0);
        for (int i = 0; i < colid; i++)
            br.BaseStream.Seek(DbaseColumns[i].Length, SeekOrigin.Current);
        return ReadDbfValue(DbaseColumns[colid]);
    }
```

7.3.3.11　ParseDbfHeader（string filename）

转换属性文件的文件头，首先调用 ReadByte 方法，读取 DBF 的类型，若类型的值不是 0X03，则抛出异常，提示不支持此类型 DBF 文件。接着读取数据最后更新时间、文件头的长度、记录的个数、文件的编码方式、属性列的个数，代码如下：

```
private void ParseDbfHeader(string filename)
{
        if (br.ReadByte() != 0x03)
            throw new NotSupportedException("Unsupported DBF Type");
        _lastUpdate = new DateTime((int)br.ReadByte() + 1900, (int)br.ReadByte(), (int)br.ReadByte());
        //Read the last update date
        _NumberOfRecords = br.ReadInt32(); // read number of records.
        _HeaderLength = br.ReadInt16(); // read length of header structure.
        _RecordLength = br.ReadInt16(); // read length of a record
        fs.Seek(29, SeekOrigin.Begin); //Seek to encoding flag
        _FileEncoding = GetDbaseLanguageDriver(br.ReadByte()); //Read and parse Language driver
        fs.Seek(32, SeekOrigin.Begin); //Move past the reserved bytes
        int NumberOfColumns = (_HeaderLength - 31)/32;
```

新建一个 DbaseField 对象，用于保存读取的属性列，这里需要遍历每个属性列，获取属性列的名称和类型，代码如下所示：

```
        DbaseColumns = new DbaseField[NumberOfColumns];
        for (int i = 0; i < DbaseColumns.Length; i++)
        {
            DbaseColumns[i] = new DbaseField();
            DbaseColumns[i].ColumnName = Encoding.UTF7.GetString((br.ReadBytes(11))).Replace("\0", "").Trim();
            char fieldtype = br.ReadChar();
            switch (fieldtype)
            {
                case 'L':
                    DbaseColumns[i].DataType = typeof(bool);
```

```
                    break;
                case 'C':
                    DbaseColumns[i].DataType = typeof(string);
                    break;
                case 'D':
                    DbaseColumns[i].DataType = typeof(DateTime);
                    break;
                case 'N':
                    DbaseColumns[i].DataType = typeof(double);
                    break;
                case 'F':
                    DbaseColumns[i].DataType = typeof(float);
                    break;
                case 'B':
                    DbaseColumns[i].DataType = typeof(byte[]);
                    break;
                default:
                    throw(new NotSupportedException("Invalid or unknown DBase field type '" +
fieldtype +"' in column '" + DbaseColumns[i].ColumnName + "'"));
        }
```

设置列的 Address 属性、长度 Length、精度 Decimals，最后设置 HeaderIsParsed 属性为真，然后调用 CreateBaseTable 创建基本的属性数据表：

```
                DbaseColumns[i].Address = br.ReadInt32();
                int Length = (int)br.ReadByte();
                if(Length < 0) Length = Length + 256;
                DbaseColumns[i].Length = Length;
                DbaseColumns[i].Decimals = (int)br.ReadByte();
                //If the double-type doesn't have any decimals, make the type an integer
                if(DbaseColumns[i].Decimals == 0 && DbaseColumns[i].DataType == typeof(double))
                    if(DbaseColumns[i].Length <= 2)
                        DbaseColumns[i].DataType = typeof(Int16);
                    else if(DbaseColumns[i].Length <= 4)
                        DbaseColumns[i].DataType = typeof(Int32);
                    else
                        DbaseColumns[i].DataType = typeof(Int64);
                fs.Seek(fs.Position + 14, 0);
            }
            HeaderIsParsed = true;
            CreateBaseTable();
        }
```

7.3.3.12 ReadDbfValue (DbaseField dbf)

读取属性值，代码如下：

```csharp
private object ReadDbfValue(DbaseField dbf)
{
    switch(dbf.DataType.ToString())
    {
        case "System.String":
            if(_Encoding == null)
                return _FileEncoding.GetString(br.ReadBytes(dbf.Length)).Replace("\0","").Trim();
            else
                return _Encoding.GetString(br.ReadBytes(dbf.Length)).Replace("\0","").Trim();
        case "System.Double":
            string temp = Encoding.UTF7.GetString(br.ReadBytes(dbf.Length)).Replace("\0","").Trim();
            double dbl = 0;
            if(double.TryParse(temp, NumberStyles.Float, Map.NumberFormatEnUs, out dbl))
                return dbl;
            else return DBNull.Value;
        case "System.Int16":
            string temp16 = Encoding.UTF7.GetString((br.ReadBytes(dbf.Length))).Replace("\0","").Trim();
            nt16 i16 = 0;
            if(Int16.TryParse(temp16, NumberStyles.Float, Map.NumberFormatEnUs, out i16))
                return i16;
            else return DBNull.Value;
        case "System.Int32":
            string temp32 = Encoding.UTF7.GetString((br.ReadBytes(dbf.Length))).Replace("\0","").Trim();
            Int32 i32 = 0;
            if(Int32.TryParse(temp32, NumberStyles.Float, Map.NumberFormatEnUs, out i32))
                return i32;
            else return DBNull.Value;
        case "System.Int64":
            string temp64 = Encoding.UTF7.GetString((br.ReadBytes(dbf.Length))).Replace("\0","").Trim();
            Int64 i64 = 0;
            if(Int64.TryParse(temp64, NumberStyles.Float, Map.NumberFormatEnUs, out i64))
                return i64;
            else return DBNull.Value;
        case "System.Single":
            string temp4 = Encoding.UTF8.GetString((br.ReadBytes(dbf.Length)));
            float f = 0;
```

```
            if (float.TryParse(temp4, NumberStyles.Float, Map.NumberFormatEnUs, out f))
                return f;
            else return DBNull.Value;
    case "System.Boolean":
        char tempChar = br.ReadChar();
        return ((tempChar == 'T') || (tempChar == 't') || (tempChar == 'Y') || (tempChar == 'y'));
    case "System.DateTime":
        DateTime date;
#if ! MONO
        if(DateTime.TryParseExact(Encoding.UTF7.GetString((br.ReadBytes(8))),
            "yyyyMMdd", Map.NumberFormatEnUs, DateTimeStyles.None, out date)) return date;
        Else return DBNull.Value;
#else
        try
        {
            return date = DateTime.ParseExact(System.Text.Encoding.UTF7.GetString((br.ReadBytes(8))),
            "yyyyMMdd",SharpMap.Map.numberFormat_EnUS,System.Globalization.DateTimeStyles.None );
        }
        catch ( Exception e )
        {return DBNull.Value; }
#endif
    default:
        throw (new NotSupportedException("Cannot parse DBase field '" + dbf.ColumnName + "' of type '" + dbf.DataType.ToString() + "'"));
    }
}
```

7.4 数据提供者 ShapeFile

Shapefiles 是 ESRI 提供的一种矢量数据格式，它没有拓扑信息，一个 Shapefiles 由一组文件组成，其中必要的基本文件包括坐标文件（.shp）、索引文件（.shx）和属性文件（.dbf）三个文件。

坐标文件（.shp）用于记录空间坐标信息。它由头文件和实体信息两部分构成。坐标文件的文件头是一个长度固定（100 bytes）的记录段，一共有 9 个 int 型和 7 个 double 型数据。

实体信息负责记录坐标信息，它以记录段为基本单位，每一个记录段记录一个地理实体目标的坐标信息，每个记录段分为记录头和记录内容两部分。

记录头的内容包括记录号（Record Number）和坐标记录长度（Content Length）两个记录项。它们的位序都是 big。记录号（Record Number）和坐标记录长度（Content Length）两个记录项都是 int 型，并且 ShapeFile 文件中的记录号都是从 1 开始的。

记录内容包括目标的几何类型（ShapeType）和具体的坐标记录（X、Y），记录内容因要素几何类型的不同其具体的内容及格式都有所不同。下面分别介绍点状目标（Point）、线状目标（PolyLine）和面状目标（Polygon）三种几何类型的 .shp 文件的记录内容。

属性文件（.dbf）用于记录属性信息。它是一个标准的 DBF 文件，也是由头文件和实体信息两部分构成的。

索引文件（.shx）主要包含坐标文件的索引信息，文件中每个记录包含对应的坐标文件记录距离坐标文件的文件头的偏移量。通过索引文件可以很方便地在坐标文件中定位到指定目标的坐标信息。

索引文件也是由头文件和实体信息两部分构成的，其中文件头部分是一个长度固定（100 bytes）的记录段，其内容与坐标文件的文件头基本一致。它的实体信息以记录为基本单位，每一条记录包括偏移量（Offset）和记录段长度（Content Length）两个记录项，它们的位序都是 big，两个记录项都是 int 型。

7.4.1 ShapeFile 的数据成员

7.4.1.1 _CoordinateSystem
坐标系统。SharpMap 直接使用开源类库 Proj4 中定义的坐标系统，代码如下：

private ICoordinateSystem _CoordinateSystem;

7.4.1.2 _CoordsysReadFromFile
指示坐标系统是否从文件中读取，代码如下：

private bool _CoordsysReadFromFile = false;

7.4.1.3 _Envelope
表示 shapeFile 的数据集的外包矩形框，代码如下：

private BoundingBox _Envelope;

7.4.1.4 _FeatureCount
ShapeFile 数据集中的要素数量，代码如下：

private int _FeatureCount;

7.4.1.5 _FileBasedIndex
是否有基于文件的索引，代码如下：

private bool _FileBasedIndex;

7.4.1.6 _Filename
ShapeFile 文件的文件名，代码如下：

private string _Filename;

7.4.1.7 _FilterDelegate

用于过滤要素的过滤器委托，代码如下：

```
public delegate bool FilterMethod(FeatureDataRow dr);
private FilterMethod _FilterDelegate;
```

7.4.1.8 _IsOpen

指示 ShapeFile 数据源是否被打开，代码如下：

```
private bool _IsOpen;
```

7.4.1.9 _ShapeType

Shape 的类型，有点、线、多边形等，代码如下：

```
private ShapeType _ShapeType;
```

7.4.1.10 _SRID

空间参考的 ID 编号，代码如下：

```
private int _SRID = -1;
```

7.4.1.11 brShapeFile

用二进制数据读取 ShapeFile 文件，它使用的是 .NET 的 BinaryReader，以流的形式读取 ShapeFile 的 .shp 文件（坐标信息文件），代码如下：

```
private BinaryReader brShapeFile;
```

7.4.1.12 brShapeIndex

用二进制数据读取 ShapeFile 文件，它使用的是 .NET 的 BinaryReader，以流的形式读取 ShapeFile 的 .shx 文件（索引信息文件），代码如下：

```
private BinaryReader brShapeIndex;
```

7.4.1.13 dbaseFile

用 DbaseReader 读取 ShapeFile 的 .dbf 文件（属性信息文件），代码如下：

```
private DbaseReader dbaseFile;
```

7.4.1.14 fsShapeFile

用文件流的方式读取坐标信息文件，代码如下：

```
private FileStream fsShapeFile;
```

7.4.1.15 fsShapeIndex

读取 ShapeFile 索引文件 .shx 的文件流，代码如下：

```
private FileStream fsShapeIndex;
```

7.4.1.16 tree

基于二叉树的空间索引，类型为 SharpMap.Utilities.SpatialIndexing 空间下的 QuadTree

类(由于篇幅关系,请读者自行研究),代码如下:

```
private QuadTree tree;
```

7.4.1.17 disposed
表示资源是否已经被系统回收处理,代码如下:

```
private bool disposed = false;
```

7.4.2 ShapeFile 的属性

7.4.2.1 ConnectionID
数据源的 ID 编号,这个编号对连接的共享非常重要,代码如下:

```
public string ConnectionID
{
    get { return _Filename; }
}
```

7.4.2.2 SRID
获取和设置空间参考的 ID 编号,代码如下:

```
public int SRID
{
    get { return _SRID; }
    set { _SRID = value; }
}
```

7.4.2.3 IsOpen
数据源是否被打开,代码如下:

```
public bool IsOpen
{
    get { return _IsOpen; }
}
```

7.4.2.4 CoordinateSystem
获取和设置数据源的坐标系统。从其 set 访问器方法中可以看出,目前 SharpMap 并不支持从文件中读取坐标系统,代码如下:

```
public ICoordinateSystem CoordinateSystem
{
    get { return _CoordinateSystem; }
    set
    {
        if (_CoordsysReadFromFile)
            throw new ApplicationException("Coordinate system is specified in projection file and is
```

7.4 数据提供者 ShapeFile

```
read only");
        _CoordinateSystem = value;
    }
}
```

7.4.2.5 ShapeType

获取几何形状的类型,这是 SharpMap 封装的一个枚举值,代码如下:

```
public ShapeType ShapeType
{
    get { return _ShapeType; }
}
```

7.4.2.6 Filename

获取和设置 ShapeFile 文件名称。在其 set 访问器方法中,若名称不与现有名称相同,则设置_Filename 为要设置的值,若 IsOpen 为真,则抛出异常,提示不能在数据源打开的情况下更改文件名称。接着调用 ParseHeader 转换文件头,调用 ParseProjection 方法转换投影,并设置 tree 为空,代码如下:

```
public string Filename
{
    get { return _Filename; }
    set
    {
        if (value != _Filename)
        {
            _Filename = value;
            if (IsOpen)
                throw new ApplicationException("Cannot change filename while datasource is open");
            ParseHeader();
            ParseProjection();
            tree = null;
        }
    }
}
```

7.4.2.7 Encoding

获取和设置 Dbase 类型的 DBF 文件的编码方式,代码如下:

```
public Encoding Encoding
{
    get { return dbaseFile.Encoding; }
    set { dbaseFile.Encoding = value; }
}
```

7.4.2.8 FilterDelegate

数据过滤器委托,代码如下:

```
public FilterMethod
{
    get { return _FilterDelegate; }
    set { _FilterDelegate = value; }
}
```

7.4.3 ShapeFile 的方法

7.4.3.1 ShapeFile(string filename, bool fileBasedIndex)

ShapeFile 数据源的构造方法,它有两个参数,一为 ShapeFile 的文件名称 filename,二为是否基于文件索引 fileBasedIndex。这里首先设置_Filename 的值为 filename;然后判断索引文件是否存在,并设置_FileBasedIndex 值;新建一个 DbaseReader 对象,用于读取属性数据,接着调用 ParseHeader 解析文件头,调用 ParseProjection 方法解析投影,代码如下:

```
public ShapeFile(string filename, bool fileBasedIndex)
{
    _Filename = filename;
    _FileBasedIndex = (fileBasedIndex) && File.Exists(Path.ChangeExtension(filename,".shx"));
    if (File.Exists(dbffile)) dbaseFile = new DbaseReader(dbffile);
    ParseHeader();
    ParseProjection();
}
```

7.4.3.2 ShapeFile(string filename)

构造函数的重载。这里仅需传入文件路径参数,它默认设置基于文件索引为 false,调用上面的重载版本来实例化 ShapeFile 数据源,代码如下:

```
public ShapeFile(string filename) : this(filename, false)
{
}
```

7.4.3.3 Dispose(bool disposing)

实现 Dispose 设计模式的方法,负责资源清理工作,代码如下:

```
private void Dispose(bool disposing)
{
    if (!disposed)
    {
        if (disposing)
        {
            Close();
            _Envelope = null;
```

```
                    tree = null;
                }
                disposed = true;
            }
        }
```

7.4.3.4 Dispose()

实现 Dispose 设计模式的方法，负责资源清理工作，代码如下：

```
        public void Dispose()
        {
            Dispose(true);
            GC.SuppressFinalize(this);
        }
```

7.4.3.5 ~ShapeFile()

析构函数，执行垃圾清理，在析构函数内部调用 Dispsoe 方法，保证资源能够及时释放，代码如下：

```
        ~ShapeFile()
        {
            Dispose();
        }
```

7.4.3.6 Close()

关闭数据源，在关闭数据源同时也关闭所有的数据读取流，代码如下：

```
        public void Close()
        {
            if(! disposed)
            {
                if(_IsOpen)
                {
                    brShapeFile.Close();
                    fsShapeFile.Close();
                    brShapeIndex.Close();
                    fsShapeIndex.Close();
                    if(dbaseFile ! = null)
                        dbaseFile.Close();
                    _IsOpen = false;
                }
            }
        }
```

7.4.3.7 Open()

打开数据源，初始化索引信息文件读取流 fsShapeIndex、索引信息文件二进制读取流

brShapeIndex、坐标信息文件流 fsShapeFile、二进制坐标信息流 brShapeFile，然后调用 InitializeShape 方法初始化索引信息，若 dbaseFile 不为空，调用其 Open 方法初始化属性信息，最后设置_IsOpen 值为 true，代码如下：

```
public void Open()
{
    if (!_IsOpen)
    {
        fsShapeIndex = new FileStream(_Filename.Remove(_Filename.Length - 4, 4) + ".shx",
            FileMode.Open, FileAccess.Read);
        brShapeIndex = new BinaryReader(fsShapeIndex, Encoding.Unicode);
        fsShapeFile = new FileStream(_Filename, FileMode.Open, FileAccess.Read);
        brShapeFile = new BinaryReader(fsShapeFile);
        InitializeShape(_Filename, _FileBasedIndex);
        if (dbaseFile != null)
            dbaseFile.Open();
        _IsOpen = true;
    }
}
```

7.4.3.8　GetGeometriesInView（BoundingBox bbox）

获取指定外包框中的几何对象集合。在这个方法的内部，首先调用 GetObjectIDsInView() 方法，获取在此外包框中的几何对象的 ID 编号集合；接着遍历这个包含 ID 编号的集合，通过几何对象的 ID 得到每个几何对象，这个过程使用的是 GetGeometryByID 方法。值得注意的是，这个方法本身不保证每个要素精确地在这个指定的外包框中，只是确保要素的外包矩形框在指定的外包框中，代码如下：

```
public Collection<Geometry> GetGeometriesInView(BoundingBox bbox)
{
    Collection<uint> objectlist = GetObjectIDsInView(bbox);
    if (objectlist.Count == 0) //no features found. Return an empty set
        return new Collection<Geometry>();
    Collection<Geometry> geometries = new Collection<Geometry>();
    for (int i = 0; i < objectlist.Count; i++)
    {
        Geometry g = GetGeometryByID(objectlist[i]);
        if (g != null)
            geometries.Add(g);
    }
    return geometries;
}
```

7.4.3.9　ExecuteIntersectionQuery（BoundingBox bbox，FeatureDataSet ds）

执行相交查询，这是实现 IProvider 接口的方法。先得到在指定外包框中的对象的 ID

编号的集合，然后通过 ID 编号得到这个要素，将其转换为 FeatureDataRow，并将转换后的 FeatureDataRow 加入到 FeatureDataTable 中，最后将其加入传入的参数 FeatureDataSet 中，代码如下：

```
public void ExecuteIntersectionQuery(BoundingBox bbox, FeatureDataSet ds)
{
    Collection<uint> objectlist = GetObjectIDsInView(bbox);
    FeatureDataTable dt = dbaseFile.NewTable;
    for (int i = 0; i < objectlist.Count; i++)
    {
        FeatureDataRow fdr = GetFeature(objectlist[i], dt);
        if (fdr != null) dt.AddRow(fdr);
    }
    ds.Tables.Add(dt);
}
```

7.4.3.10 GetObjectIDsInView (BoundingBox bbox)

得到指定外包框中的要素的 ID 编号集合。数据源未打开，则抛出异常；最后调用空间参考树的 Search() 方法进行查询，代码如下：

```
public Collection<uint> GetObjectIDsInView(BoundingBox bbox)
{
    if (!IsOpen)
        throw (new ApplicationException("An attempt was made to read from a closed datasource"));
    return tree.Search(bbox);
}
```

7.4.3.11 GetGeometryByID (uint oid)

通过几何对象的 ID 编号得到这个几何对象。首先检查过滤器委托是否为空，不为空则调用 GetFeature() 方法得到几何对象；若为空则调用 ReadGeometry() 方法读取几何对象，代码如下：

```
public Geometry GetGeometryByID(uint oid)
{
    if (FilterDelegate != null) //Apply filtering
    {
        FeatureDataRow fdr = GetFeature(oid);
        if (fdr != null)
            return fdr.Geometry;
        return null;
    }
    return ReadGeometry(oid);
}
```

7.4.3.12 ExecuteIntersectionQuery (Geometry geom, FeatureDataSet ds)

执行相交查询,这是一个重载方法。它首先得到给定要素的外包框矩形,然后直接调用上面的重载版本的方法执行相交查询,代码如下:

```
public void ExecuteIntersectionQuery(Geometry geom, FeatureDataSet ds)
{
    FeatureDataTable dt = dbaseFile.NewTable;
    BoundingBox bbox = geom.GetBoundingBox();
    ExecuteIntersectionQuery(bbox, ds);
}
```

7.4.3.13 GetFeatureCount ( )

得到要素的数目,它直接返回_FeatureCount 的值,代码如下:

```
public int GetFeatureCount()
{
    return _FeatureCount;
}
```

7.4.3.14 GetFeature (uint RowID)

通过行编号得到要素的数据行,它在内部调用 GetFeature 的另一个重载版本,代码如下:

```
public FeatureDataRow GetFeature(uint RowID)
{
    return GetFeature(RowID, dbaseFile.NewTable);
}
```

7.4.3.15 GetExtents ( )

得到所有要素的外包框的交集作为 ShapeFile 数据源的外包框矩形。若空间索引树为空,则抛出异常,提示文件未建立空间索引,若不为空,直接返回空间索引树的 Box 属性,代码如下:

```
public BoundingBox GetExtents()
{
    if (tree == null)
        throw new ApplicationException("File hasn't been spatially indexed. Try opening the datasource before retriving extents");
    return tree.Box;
}
```

7.4.3.16 InitializeShape (string filename, bool fileBasedIndex)

初始化 ShapeFile。这里检查文件路径的合法性和文件的合法性,然后调用 LoadSpatialIndex 方法生成空间索引树,代码如下:

```
private void InitializeShape(string filename, bool fileBasedIndex)
```

7.4 数据提供者 ShapeFile

```
    if (! File.Exists(filename))
        throw new FileNotFoundException(String.Format("Could not find file \"{0}\"", filename));
    if (! filename.ToLower().EndsWith(".shp"))
        throw (new Exception("Invalid shapefile filename: " + filename));
    LoadSpatialIndex(fileBasedIndex); //Load spatial index
}
```

7.4.3.17 ParseHeader()

解析 ShapeFile 文件的索引文件 .shx 的头部。以二进制文件流的方式读取并转换得到要素的个数 _FeatureCount、要素的集合类型 _ShapeType、外包框 _Envelope,代码如下:

```
private void ParseHeader()
{
    fsShapeIndex = new FileStream(Path.ChangeExtension(_Filename, ".shx"), FileMode.Open, FileAccess.Read);
    brShapeIndex = new BinaryReader(fsShapeIndex, Encoding.Unicode);
    brShapeIndex.BaseStream.Seek(0, 0);
    if (brShapeIndex.ReadInt32() ! = 170328064)
        //File Code is actually 9994, but in Little Endian Byte Order this is '170328064'
        //throw (new ApplicationException("Invalid Shapefile Index (.shx)"));
    brShapeIndex.BaseStream.Seek(24, 0); //seek to File Length
    int IndexFileSize = SwapByteOrder(brShapeIndex.ReadInt32());
        //Read filelength as big - endian. The length is based on 16bit words
    _FeatureCount = (2 * IndexFileSize - 100)/8;
        //Calculate FeatureCount. Each feature takes up 8 bytes. The header is 100 bytes
    brShapeIndex.BaseStream.Seek(32, 0); //seek to ShapeType
    _ShapeType = (ShapeType) brShapeIndex.ReadInt32();
    //Read the spatial bounding box of the contents
    brShapeIndex.BaseStream.Seek(36, 0); //seek to box
    _Envelope = new BoundingBox(brShapeIndex.ReadDouble(), brShapeIndex.ReadDouble(),
        brShapeIndex.ReadDouble(), brShapeIndex.ReadDouble());
    brShapeIndex.Close();
    fsShapeIndex.Close();
}
```

7.4.3.18 ParseProjection()

解析投影。首先读取 ShapeFile 文件的投影文件 .prj,调用 Proj4 类库的 CoordinateSystemWktReader 方法转换投影,并设置读取后的 ShapeFile 数据源的投影为转换后的投影,代码如下:

```
private void ParseProjection()
{
    String projfile = Path.GetDirectoryName(Filename) + "\\" + Path.GetFileName
```

```csharp
          WithoutExtension(Filename) + ".prj";
            if (File.Exists(projfile))
            {
                try
                {
                    string wkt = File.ReadAllText(projfile);
                    _CoordinateSystem = (ICoordinateSystem)CoordinateSystemWktReader.Parse(wkt);
                    _CoordsysReadFromFile = true;
                }
                catch (Exception ex)
                {
                        Trace.TraceWarning("Coordinate system file '" + projfile + "' found, but could not be parsed. WKT parser returned:" + ex.Message);
                        throw (ex);
                }
            }
        }
```

7.4.3.19 ReadIndex ( )

读取 ShapeFile 文件的索引文件 .shx，返回坐标偏移量的集合。这里首先实例化一个大小为要素数量的数组，然后遍历每个要素读取每个要素（实体信息）的偏移量和记录段长度，将偏移量保存于数组中并返回，代码如下：

```csharp
        private int[] ReadIndex()
        {
            int[] OffsetOfRecord = new int[_FeatureCount];
            brShapeIndex.BaseStream.Seek(100, 0); //skip the header
            for (int x = 0; x < _FeatureCount; ++x)
            {
                OffsetOfRecord[x] = 2 * SwapByteOrder(brShapeIndex.ReadInt32());
                brShapeIndex.BaseStream.Seek(brShapeIndex.BaseStream.Position + 4, 0);
            }
            return OffsetOfRecord;
        }
```

7.4.3.20 GetShapeIndex (uint n)

获取 ShapeFile 几何数据的地址。这里首先调用索引信息二进制读取对象 brShapeIndex 的 Seek 方法，直接定位到索引信息的地址，然后返回读取的数据，代码如下：

```csharp
        private int GetShapeIndex(uint n)
        {
            brShapeIndex.BaseStream.Seek(100 + n * 8, 0); //seek to the position of the index
            return 2 * SwapByteOrder(brShapeIndex.ReadInt32()); //Read shape data position
        }
```

7.4.3.21 SwapByteOrder (int i)

此方法用于反转位序,ShapeFile 文件的位序有 Little 和 big 之分,两者的区别在于它们二进制存储的位序相反。通常,数据的位序都是 Little,但在有些情况下可能会是 big。对于位序是 big 的数据在读取时要进行转换。转换原理非常简单,就是交换字节顺序,如下所示,调用 Array.Reverse 方法进行位序反转,代码如下:

```
private int SwapByteOrder(int i)
{
    byte[] buffer = BitConverter.GetBytes(i);
    Array.Reverse(buffer, 0, buffer.Length);
    return BitConverter.ToInt32(buffer, 0);
}
```

7.4.3.22 CreateSpatialIndexFromFile (string filename)

从指定的文件路径中创建空间索引树,对于存在空间索引文件的 Shapefile 文件我们直接读取后创建空间索引树;对于没有空间索引文件的 ShapeFile 文件,我们先创建空间索引树,然后以文件的形式保存此空间索引,代码如下:

```
private QuadTree CreateSpatialIndexFromFile(string filename)
{
    if (File.Exists(filename + ".sidx"))
    {
        try
        {return QuadTree.FromFile(filename + ".sidx");}
        catch (QuadTree.ObsoleteFileFormatException)
        {
            File.Delete(filename + ".sidx");
            return CreateSpatialIndexFromFile(filename);
        }
        catch (Exception ex)
        {throw ex;}
    }
    else
    {
        QuadTree tree = CreateSpatialIndex(_Filename);
        tree.SaveIndex(filename + ".sidx");
        return tree;
    }
}
```

7.4.3.23 CreateSpatialIndex (string filename)

创建指定 ShapeFile 文件的空间索引树,这些树种保存着几何对象的外包框。首先新建一个类型为 List < QuadTree.BoxObjects > 的几何对象 objList,用于保存所有要素的外包框矩形；然后调用 GetAllFeatureBoundingBoxes 方法获取所有的外包框矩形，并对其进行遍历，将边界值合法的外包框矩形加入到 objList 中；新建一个 Heuristic 类型的结构体变量，设置最大深度 maxdepth，值为以 2 为底要素个数的对数值取整、最小误差 minerror 为 10、节点上目标的个数 tartricnt 为 5、节点上最小对象的个数 mintricnt 为 2，然后将其作为参数新建 QuadTree 类型的对象并返回，代码如下：

```
private QuadTree CreateSpatialIndex(string filename)
{
    List<QuadTree.BoxObjects> objList = new List<QuadTree.BoxObjects>();
    //Convert all the geometries to boundingboxes
    uint i = 0;
    foreach (BoundingBox box in GetAllFeatureBoundingBoxes())
    {
        if (! double.IsNaN(box.Left) && ! double.IsNaN(box.Right) && ! double.IsNaN(box.Bottom) &&
            ! double.IsNaN(box.Top))
        {
            QuadTree.BoxObjects g = new QuadTree.BoxObjects();
            g.box = box;
            g.ID = i;
            objList.Add(g);
            i++;
        }
    }
    Heuristic heur;
    heur.maxdepth = (int) Math.Ceiling(Math.Log(GetFeatureCount(), 2));
    heur.minerror = 10;
    heur.tartricnt = 5;
    heur.mintricnt = 2;
    return new QuadTree(objList, 0, heur);
}
```

7.4.3.24　LoadSpatialIndex()

加载空间索引，此方法设置参数都为 false，调用另一重载版本的方法，代码如下：

```
private void LoadSpatialIndex()
{
    LoadSpatialIndex(false, false);
}
```

7.4.3.25　LoadSpatialIndex(bool LoadFromFile)

是否从文件中加载空间索引，此方法也调用另一重载方法，代码如下：

```
private void LoadSpatialIndex(bool LoadFromFile)
```

```
        LoadSpatialIndex(false, LoadFromFile);
}
```

7.4.3.26 LoadSpatialIndex (bool ForceRebuild, bool LoadFromFile)

加载空间索引，参数为是否强制重建空间索引 ForseRebuild、是否从文件中读取 LoadFromFile。若空间索引树为空或者 ForceRebuild 为真，才继续操作，代码如下：

```
private void LoadSpatialIndex(bool ForceRebuild, bool LoadFromFile)
{
    if (tree == null || ForceRebuild)
    {
        // Is this a web application? If so lets store the index in the cache so we don't
        // need to rebuild it for each request
```

若 HttpContext.Current 不为空，则为 Web 应用程序，此时需要将空间索引保存在缓存中，以防在每次请求时都重建空间索引。若缓存不为空，直接调用 HttpContext.Current.Cache 的索引方法获取缓存，并将其强制转换为 QuadTree 类型，赋予 tree；否则，若 LoadFromFile 为假，调用 CreateSpatialIndex 方法创建空间索引，将结果赋予 tree，若 LoadFromFile 为真，调用 CreateSpatialIndexFromFile 方法从文件创建空间索引；最后将其插入缓存中。代码如下：

```
        if (HttpContext.Current != null)
        {
            //Check if the tree exists in the cache
            if (HttpContext.Current.Cache[_Filename] != null)
                tree = (QuadTree) HttpContext.Current.Cache[_Filename];
            else
            {
                if (!LoadFromFile) tree = CreateSpatialIndex(_Filename);
                else tree = CreateSpatialIndexFromFile(_Filename);
                HttpContext.Current.Cache.Insert(_Filename, tree, null, Cache.NoAbsoluteExpiration,TimeSpan.FromDays(1));
            }
        }
```

若不是 Web 应用程序，若 LoadFromFile 为真，调用 CreateSpatialIndexFromFile 方法从文件创建空间索引，并将结果赋予 tree：

```
        else if (!LoadFromFile)
            tree = CreateSpatialIndex(_Filename);
        else
            tree = CreateSpatialIndexFromFile(_Filename);
    }
}
```

7.4.3.27 RebuildSpatialIndex()

重建空间索引树。若 ShapeFile 数据源使用了基于文件的索引,若存在 .sidx 文件,则删除文件后调用 CreateSpatialIndexFromFile 方法从文件创建空间索引,赋予 tree 对象;否则,调用 CreateSpatialIndex 方法创建空间索引,赋予 tree 对象;若 HttpContext.Current 不为空,还要将其插入缓存中,代码如下:

```csharp
public void RebuildSpatialIndex()
{
    if(_FileBasedIndex)
    {
        if(File.Exists(_Filename + ".sidx"))
            File.Delete(_Filename + ".sidx");
        tree = CreateSpatialIndexFromFile(_Filename);
    }
    else
        tree = CreateSpatialIndex(_Filename);
    if(HttpContext.Current != null)
    {
        //TODO: Remove this when connection pooling is implemented:
        HttpContext.Current.Cache.Insert(_Filename, tree, null, Cache.NoAbsoluteExpiration, TimeSpan.FromDays(1));
    }
}
```

7.4.3.28 GetAllFeatureBoundingBoxes()

获取所有要素的外包矩形框的迭代器对象,这对建立空间索引十分有用。首先调用 ReadIndex 方法读取所有要素的偏移值,保存于整形数组 offsetOfRecord 中;若要素类型为点,遍历每个要素,读取点的 X、Y 坐标值,并将其作为参数,新建一个 BoundingBox 对象,迭代返回,代码如下:

```csharp
private IEnumerable<BoundingBox> GetAllFeatureBoundingBoxes()
{
    int[] offsetOfRecord = ReadIndex(); //Read the whole .idx file
    //List<BoundingBox> boxes = new List<BoundingBox>();
    if(_ShapeType == ShapeType.Point)
    {
        for(int a = 0; a < _FeatureCount; ++a)
        {
            fsShapeFile.Seek(offsetOfRecord[a] + 8, 0);
            if(((ShapeType)brShapeFile.ReadInt32()) != ShapeType.Null)
            {
                double x = brShapeFile.ReadDouble();
                double y = brShapeFile.ReadDouble();
                yield return new BoundingBox(x, y, x, y);
            }
        }
```

7.4 数据提供者 ShapeFile

```
        }
    }
```

若类型不为点，遍历每个要素，读取边界值，并将其作为参数，新建 BoundingBox 对象迭代返回。代码如下：

```
else
{
    for (int a = 0; a < _FeatureCount; ++a)
    {
        fsShapeFile.Seek(offsetOfRecord[a] + 8, 0);
        if ((ShapeType)brShapeFile.ReadInt32() != ShapeType.Null)
            yield return new BoundingBox(brShapeFile.ReadDouble(), brShapeFile.ReadDouble(),
                brShapeFile.ReadDouble(), brShapeFile.ReadDouble());
    }
}
```

7.4.3.29 ReadGeometry (uint oid)

通过几何对象的编号，读取几何对象。首先得到几何对象的地址，然后读取其类型，若类型为空，返回 null；若几何对象的类型为点，新建一个点对象，参数为调用 brShape-File.ReadDouble() 方法读取的 X、Y 坐标值，代码如下：

```
private Geometry ReadGeometry(uint oid)
{
    brShapeFile.BaseStream.Seek(GetShapeIndex(oid) + 8, 0);
    ShapeType type = (ShapeType)brShapeFile.ReadInt32();  //Shape type
    if (type == ShapeType.Null)
        return null;
    if (_ShapeType == ShapeType.Point || _ShapeType == ShapeType.PointM || _ShapeType == ShapeType.PointZ)
    {
        Point tempFeature = new Point();
        return new Point(brShapeFile.ReadDouble(), brShapeFile.ReadDouble());
    }
```

若几何对象的类型为复合点，先得到几何对象的地址，新建一个复合点对象 feature，然后调用 brShapeFile.ReadInt32() 方法读取多点对象中点的个数，然后逐个读取每个点的坐标值，新建点对象，并将点加入多点对象 feature 中，最后返回 feature：

```
    else if (_ShapeType == ShapeType.Multipoint || _ShapeType == ShapeType.MultiPointM ||
        _ShapeType == ShapeType.MultiPointZ)
    {
        brShapeFile.BaseStream.Seek(32 + brShapeFile.BaseStream.Position, 0);
        MultiPoint feature = new MultiPoint();
```

```
    int nPoints = brShapeFile.ReadInt32();
    if (nPoints == 0) return null;
    for (int i = 0; i < nPoints; i++)
        feature.Points.Add(new Point(brShapeFile.ReadDouble(), brShapeFile.ReadDouble()));
    return feature;
}
```

若几何对象的类型为线或者多边形，首先得到几何对象的地址，读取线对象的数量，保存于整型变量 nPart 中，若数量为 0 则返回空；读取点的个数保存于整形变量 nPoints 中，新建一个大小为 nParts + 1 的数组 segments，用于保存每个组成部分的地址，然后逐个读取每部分的地址，保存于数组 segments 中；设置数组 segments 最后一个元素的值为点的个数 nPoints；若（int）_ShapeType%10 的值为3，新建一个复合线对象 mline，然后遍历每个组成部分。对于每个组成部分，新建一个线串对象 line，然后遍历每个节点，读取节点的 X、Y 坐标值，将生成的点对象加入 line 的节点集合中，然后将线串 line 加入复合线对象中。最后，若 mline 线串的个数为 1，返回此线串；若不为 1，返回 mline。代码如下：

```
else if (_ShapeType == ShapeType.PolyLine || _ShapeType == ShapeType.Polygon ||
        _ShapeType == ShapeType.PolyLineM || _ShapeType == ShapeType.PolygonM ||
        _ShapeType == ShapeType.PolyLineZ || _ShapeType == ShapeType.PolygonZ)
{
    brShapeFile.BaseStream.Seek(32 + brShapeFile.BaseStream.Position, 0);
    int nParts = brShapeFile.ReadInt32(); // get number of parts (segments)
    if (nParts == 0) return null;
    int nPoints = brShapeFile.ReadInt32(); // get number of points
    int[] segments = new int[nParts + 1];
    for (int b = 0; b < nParts; b++)
        segments[b] = brShapeFile.ReadInt32();
    segments[nParts] = nPoints;
    if ((int)_ShapeType%10 == 3)
    {
        MultiLineString mline = new MultiLineString();
        for (int LineID = 0; LineID < nParts; LineID++)
        {
            LineString line = new LineString();
            for (int i = segments[LineID]; i < segments[LineID + 1]; i++)
                line.Vertices.Add(newPoint(brShapeFile.ReadDouble(), brShapeFile.ReadDouble()));
            mline.LineStrings.Add(line);
        }
        if (mline.LineStrings.Count == 1) return mline[0];
        return mline;
    }
```

若（int）_ShapeType%10 的值不为 3，新建一个 List < LinearRing > 类型的几何对象

rings，用于保存多边形的环，然后遍历每个组成部分，读取点的 X、Y 坐标值，将其加入环的节点集合中，然后将环加入 rings 集合中；接着判断每个环是否为顺时针方向，若不是顺时针方向，将 PolygonCount 自加 1；若 PolygonCount 的值为 1，说明目前只有一个多边形，则此时新建一个多边形对象 poly，设置 ExteriorRing 属性的值为 rings[0]，然后将其余的多边形加入多边形的 InteriorRings 集合中。代码如下：

```
else //(_ShapeType == ShapeType.Polygon etc...)
{
    List<LinearRing> rings = new List<LinearRing>();
    for (int RingID = 0; RingID < nParts; RingID++)
    {
        LinearRing ring = new LinearRing();
        for (int i = segments[RingID]; i < segments[RingID + 1]; i++)
            ring.Vertices.Add(new Point(brShapeFile.ReadDouble(), brShapeFile.ReadDouble()));
        rings.Add(ring);
    }
    bool[] IsCounterClockWise = new bool[rings.Count];
    int PolygonCount = 0;
    for (int i = 0; i < rings.Count; i++)
    {
        IsCounterClockWise[i] = rings[i].IsCCW();
        if (!IsCounterClockWise[i])
            PolygonCount++;
    }
    if (PolygonCount == 1) //We only have one polygon
    {
        Polygon poly = new Polygon();
        poly.ExteriorRing = rings[0];
        if (rings.Count > 1)
            for (int i = 1; i < rings.Count; i++)
                poly.InteriorRings.Add(rings[i]);
        return poly;
    }
```

若几何对象的类型为复合多边形，新建一个复合多边形对象 mpoly、一个多边形对象 poly，设置 poly.ExteriorRing 值为第一个环 rings[0]，然后遍历剩余的环，若环不是顺时针方向，将多边形加入 mpoly.Polygons 集合中，新建一个 Polygon 对象，参数为当前环对象，并赋予 poly；否则直接将当前环加入 poly.InteriorRings 集合中，最后将 poly 加入 mpoly.Polygons 集合中，并返回复合多边形对象 mpoly：

```
    else
    {
        MultiPolygon mpoly = new MultiPolygon();
```

```
            Polygon poly = new Polygon();
            poly.ExteriorRing = rings[0];
            for (int i = 1; i < rings.Count; i++)
            {
                if (!IsCounterClockWise[i])
                {
                    mpoly.Polygons.Add(poly);
                    poly = new Polygon(rings[i]);
                }
                else poly.InteriorRings.Add(rings[i]);
            }
            mpoly.Polygons.Add(poly);
            return mpoly;
        }
    }
}
```

其他情况下,抛出异常,提示不支持解析此类型的几何对象:

```
else
    throw(new ApplicationException("Shapefile type " + _ShapeType.ToString() + " not supported"));
}
```

7.4.3.30 GetFeature(uint RowID, FeatureDataTable dt)

在数据源中根据要素的数据行编号得到此数据行。若 dbaseFile 不为空,调用其 GetFeature 方法获取要素数据行,保存于要素数据行变量 dr 中,设置要素数据行 dr.Geometry 属性值为 ReadGeometry 方法的返回值,若过滤器委托 FilterDelegate 不为空,调用此委托后,返回 dr;若 dbaseFile 为空,则抛出异常,提示读取.DBF 文件时发生错误,代码如下:

```
public FeatureDataRow GetFeature(uint RowID, FeatureDataTable dt)
{
    if (dbaseFile != null)
    {
        FeatureDataRow dr = dbaseFile.GetFeature(RowID, dt);
        dr.Geometry = ReadGeometry(RowID);
        if (FilterDelegate == null || FilterDelegate(dr)) return dr;
        else return null;
    }
    else
        throw(new ApplicationException("An attempt was made to read DBase data from a shapefile without a valid .DBF file"));
}
```

7.5 MsSql 类

MsSql 是使用微软 SQL Sever 数据库存储空间数据的包装类。

7.5.1 MsSql 类的数据成员

7.5.1.1 _ObjectIdColumn

表示存储对象 ID 编号列的名称,代码如下:

```
private string _ObjectIdColumn;
```

7.5.1.2 disposed

资源是否已经被回收,代码如下:

```
private bool disposed = false;
```

7.5.2 MsSql 类的属性

7.5.2.1 SRID

空间参考的 ID 编号,代码如下:

```
private int _srid = -2;
public int SRID
{
    get { return _srid; }
    set { _srid = value; }
}
```

7.5.2.2 IsOpen

数据库的连接是否被打开,代码如下:

```
private bool _IsOpen;
public bool IsOpen
{
    get { return _IsOpen; }
}
```

7.5.2.3 ConnectionID

数据连接的 ID 编号,这个对于连接池非常有用,代码如下:

```
public string ConnectionID
{
    get { return _ConnectionString; }
}
```

7.5.2.4 ConnectionString

连接 Microsoft SQL Server 数据库的连接字符串，代码如下：

```
private string _ConnectionString;
public string ConnectionString
{
    get { return _ConnectionString; }
    set { _ConnectionString = value; }
}
```

7.5.2.5 Table

存储要素数据的表的名称，代码如下：

```
private string _Table;
public string Table
{
    get { return _Table; }
    set { _Table = value; }
}
```

7.5.2.6 GeometryColumn

存储要素几何信息的列名称，代码如下：

```
private string _GeometryColumn;
public string GeometryColumn
{
    get { return _GeometryColumn; }
    set { _GeometryColumn = value; }
}
```

7.5.2.7 ObjectIdColumn

对象 ID 编号列的名称，代码如下：

```
private string _ObjectIdColumn;
public string ObjectIdColumn
{
    get { return _ObjectIdColumn; }
    set { _ObjectIdColumn = value; }
}
```

7.5.2.8 DefinitionQuery

查询表达式，代码如下：

```
private string _defintionQuery;
public string DefinitionQuery
{
```

7.5 MsSql 类

```
        get { return _defintionQuery; }
        set { _defintionQuery = value; }
    }
```

7.5.2.9 Columns

数据表中所有列的集合，可以看到目前版本中并未实现这个属性。这里只是抛出未实现的异常 NotImplementedException，代码如下：

```
public DataColumnCollection
{
    get { throw new NotImplementedException(); }
}
```

7.5.3 MsSql 类的方法

7.5.3.1 MsSql (string ConnectionStr, string tablename, string geometryColumnName, string OID_ColumnName)

MsSql 数据源的构造函数，这里初始化了连接字符串、数据表名称、存储几何对象的列的名称和对象 ID 编号的名称，代码如下：

```
public MsSql(string ConnectionStr, string tablename, string geometryColumnName, string OID_ColumnName)
{
    ConnectionString = ConnectionStr;
    Table = tablename;
    GeometryColumn = geometryColumnName; //Name of column to store geometry
    ObjectIdColumn = OID_ColumnName; //Name of object ID column
}
```

7.5.3.2 Open ( )

打开数据源。这里并不是真正打开数据源，而是将 _IsOpen 设置为 True，连接池 ConnectionPooling 会根据 _IsOpen 的值，自动完成数据库的打开工作，代码如下：

```
public void Open()
{
    _IsOpen = true;
}
```

7.5.3.3 Close ( )

关闭数据源。同样的，这里并不是真正关闭数据源，而是将 _IsOpen 设置为 False，连接池 ConnectionPooling 会完成数据库的关闭工作，代码如下：

```
public void Close()
{
```

```
            _IsOpen = false;
    }
```

7.5.3.4　GetGeometriesInView（BoundingBox bbox）

这是实现 IProvider 接口的方法，获取指定外包框矩形内的几何要素。这里先调用 GetBoxClause（）方法得到限制条件的字符串，然后生成 SQL 的查询语句进行数据库查询，然后调用 GeometryFromWKB. Parse（GeometryFromWKT 类是位于 SharpMap. Converters. WellKnownText 名字空间下的一个转换类，可以将符合 OpenGIS Simple Features Specification 规范的文本格式的空间数据转换成几何对象，由于篇幅关系，请读者自行研究）方法将查询的二进制数据转换为对应的几何要素，代码如下：

```
public Collection < Geometry > GetGeometriesInView( BoundingBox bbox)
{
    Collection < Geometry > features = new Collection < Geometry > ();
    using (SqlConnection conn = new SqlConnection(_ConnectionString))
    {
        string BoxIntersect = GetBoxClause(bbox);
        string strSQL = "SELECT " + GeometryColumn + " AS Geom ";
        strSQL + = "FROM " + Table + " WHERE ";
        strSQL + = BoxIntersect;
        if (! String. IsNullOrEmpty(_defintionQuery))
            strSQL + = " AND " + DefinitionQuery;
        using (SqlCommand command = new SqlCommand(strSQL, conn))
        {
            conn. Open();
            using (SqlDataReader dr = command. ExecuteReader())
            {
                while (dr. Read())
                {
                    if (dr[0] ! = DBNull. Value)
                    {
                        Geometry geom = GeometryFromWKB. Parse((byte[]) dr[0]);
                        if (geom ! = null)
                            features. Add(geom);
                    }
                }
            }
            conn. Close();
        }
    }
    return features;
}
```

7.5.3.5　GetGeometryByID（uint oid）

通过集合要素的 ID 编号获取此几何要素，这就意味着要从数据库存储的二进制数据

7.5 MsSql 类

转换对应的几何要素。这里将给定的几何要素的 ID 编号和数据表中的 oid 列进行比较后查找，最后的转换方法同上，也是调用 GeometryFromWKB. Parse（GeometryFromWKT 类是位于 SharpMap. Converters. WellKnownText 名称空间下的一个转换类，可以将符合 OpenGIS Simple Features Specification 规范的文本格式的空间数据转换成几何对象，由于篇幅关系，请读者自行研究）方法将查询的二进制数据转换为对应的几何要素，代码如下：

```csharp
public Geometry GetGeometryByID(uint oid)
{
    Geometry geom = null;
    using (SqlConnection conn = new SqlConnection(_ConnectionString))
    {
        string strSQL = "SELECT " + GeometryColumn + " AS Geom FROM " + Table + " WHERE " + ObjectIdColumn +
                        " = '" + oid.ToString() + "'";
        conn.Open();
        using (SqlCommand command = new SqlCommand(strSQL, conn))
        {
            using (SqlDataReader dr = command.ExecuteReader())
            {
                while (dr.Read())
                {
                    if (dr[0] != DBNull.Value)
                        geom = GeometryFromWKB.Parse((byte[])dr[0]);
                }
            }
        }
        conn.Close();
    }
    return geom;
}
```

7.5.3.6 GetObjectIDsInView（BoundingBox bbox）

获取指定外包框内的要素的 ID 编号的集合，同样的，先调用 GetBoxClause 得到查询数据库的限制条件（作为 Where 子句），然后将查询到的 oid 列对应的数据集合返回，代码如下：

```csharp
public Collection<uint> GetObjectIDsInView(BoundingBox bbox)
{
    Collection<uint> objectlist = new Collection<uint>();
    using (SqlConnection conn = new SqlConnection(_ConnectionString))
    {
        string strSQL = "SELECT " + ObjectIdColumn + " ";
        strSQL += "FROM " + Table + " WHERE ";
```

```csharp
            strSQL += GetBoxClause(bbox);
            if (!String.IsNullOrEmpty(_defintionQuery))
                strSQL += " AND " + DefinitionQuery + " AND ";
            using (SqlCommand command = new SqlCommand(strSQL, conn))
            {
                conn.Open();
                using (SqlDataReader dr = command.ExecuteReader())
                {
                    while (dr.Read())
                    {
                        if (dr[0] != DBNull.Value)
                        {
                            uint ID = (uint)(int)dr[0];
                            objectlist.Add(ID);
                        }
                    }
                }
                conn.Close();
            }
        }
        return objectlist;
    }
```

7.5.3.7 ExecuteIntersectionQuery(Geometry geom, FeatureDataSet ds)

执行相交查询,可以看到,目前的 SharpMap 中并未实现该功能。这里简单地抛出未实现异常,代码如下:

```csharp
    public void ExecuteIntersectionQuery(Geometry geom, FeatureDataSet ds)
    {
        throw new NotImplementedException();
    }
```

7.5.3.8 GetFeatureCount()

获取要素的数量。这里直接调用 SQL 的 Count 函数进行统计,返回在查询语句_ConnectionString 的限制下的要素数目,代码如下:

```csharp
    public int GetFeatureCount()
    {
        int count = 0;
        using (SqlConnection conn = new SqlConnection(_ConnectionString))
        {
            string strSQL = "SELECT COUNT(*) FROM " + Table;
            if (!String.IsNullOrEmpty(_defintionQuery))
                strSQL += " WHERE " + DefinitionQuery;
```

7.5　MsSql 类

```
        using (SqlCommand command = new SqlCommand(strSQL, conn))
        {
            conn.Open();
            count = (int)command.ExecuteScalar();
            conn.Close();
        }
    }
    return count;
}
```

7.5.3.9　GetFeature（uint RowID）

通过要素的行编号 ID 获取此数据行。这里先构造查询字符串，条件为对象 ID 列的值，与给定的 RowID 值相等，然后查询数据库，获取符合条件的数据集，接着遍历查询结果，将其转换为对应的要素数据行并返回，若查询结果为空则返回 null，代码如下：

```
public FeatureDataRow GetFeature(uint RowID)
{
    using (SqlConnection conn = new SqlConnection(_ConnectionString))
    {
        string strSQL = "SELECT *, " + GeometryColumn + " AS sharpmap_tempgeometry
                FROM " + Table + " WHERE " +
                ObjectIdColumn + " = '" + RowID.ToString() + "'";
        using (SqlDataAdapter adapter = new SqlDataAdapter(strSQL, conn))
        {
            DataSet ds = new DataSet();
            conn.Open();
            adapter.Fill(ds);
            conn.Close();
            if (ds.Tables.Count > 0)
            {
                FeatureDataTable fdt = new FeatureDataTable(ds.Tables[0]);
                foreach (DataColumn col in ds.Tables[0].Columns)
                    if (col.ColumnName != GeometryColumn && col.ColumnName != "sharpmap_tempgeometry" &&
                        !col.ColumnName.StartsWith("Envelope_"))
                        fdt.Columns.Add(col.ColumnName, col.DataType, col.Expression);
                if (ds.Tables[0].Rows.Count > 0)
                {
                    DataRow dr = ds.Tables[0].Rows[0];
                    FeatureDataRow fdr = fdt.NewRow();
                    foreach (DataColumn col in ds.Tables[0].Columns)
                        if (col.ColumnName != GeometryColumn && col.ColumnName != "sharpmap_tempgeometry" &&
```

```
                        ! col. ColumnName. StartsWith("Envelope_"))
                            fdr[col. ColumnName] = dr[col];
                        if (dr["sharpmap_tempgeometry"] ! = DBNull. Value)
                            fdr. Geometry = GeometryFromWKB. Parse((byte[]) dr["sharpmap_
tempgeometry"]);
                        return fdr;
                    }
                    else
                        return null;
                }
                else
                    return null;
            }
        }
```

7.5.3.10 GetExtents()

得到所有要素的外包框的交集,在 MsSql 数据源中,直接得到各个矩形边界的最大值组成新的外包框返回即可。如下,先在 Envelope_MinX、Envelope_MinY、Envelope_MaxX、Envelope_MaxY 四个列中分别查找最小值或最大值,将查询的结果强制转换为 float 类型,作为参数构造新的 BoundingBox 对象 box,并返回此 box:

```
public BoundingBox GetExtents()
{
    BoundingBox box = null;
    using (SqlConnection conn = new SqlConnection(_ConnectionString))
    {
        string strSQL =
            "SELECT Min(Envelope_MinX) AS MinX, Min(Envelope_MinY) AS MinY, Max
(Envelope_MaxX) AS MaxX, Max(Envelope_MaxY) AS MaxY FROM " + Table;
        if (! String. IsNullOrEmpty(_defintionQuery))
            strSQL + = " WHERE " + DefinitionQuery;
        using (SqlCommand command = new SqlCommand(strSQL, conn))
        {
            conn. Open();
            using (SqlDataReader dr = command. ExecuteReader())
                if (dr. Read())
                {
                    box = new BoundingBox((float) dr[0], (float) dr[1], (float) dr[2],
(float) dr[3]);
                }
            conn. Close();
        }
```

            return box;
        }
    }

7.5.3.11 ExecuteIntersectionQuery (BoundingBox bbox, FeatureDataSet ds)

执行相交查询,此方法过程与 GetObjectIDsInView 方法类似。这里先生成查询限制语句并执行查询,然后遍历查询结果,将结果转换为相应的要素数据行类型,将要素数据行加入要素数据表 fdt,将要素数据表加入参数要素数据集 ds 中,代码如下:

```
public void ExecuteIntersectionQuery(BoundingBox bbox, FeatureDataSet ds)
{
    //List < Geometries. Geometry > features = new List < SharpMap. Geometries. Geometry > ();
    using (SqlConnection conn = new SqlConnection(_ConnectionString))
    {
        string strSQL = "SELECT * ," + GeometryColumn + " AS sharpmap_tempgeometry ";
        strSQL + = "FROM " + Table + " WHERE ";
        strSQL + = GetBoxClause(bbox);
        if (! String. IsNullOrEmpty(_defintionQuery))
            strSQL + = " AND " + DefinitionQuery;
        using (SqlDataAdapter adapter = new SqlDataAdapter(strSQL, conn))
        {
            conn. Open();
            DataSet ds2 = new DataSet();
            adapter. Fill(ds2);
            conn. Close();
            if (ds2. Tables. Count > 0)
            {
                FeatureDataTable fdt = new FeatureDataTable(ds2. Tables[0]);
                foreach (DataColumn col in ds2. Tables[0]. Columns)
                    if (col. ColumnName ! = GeometryColumn && col. ColumnName ! = "sharpmap_tempgeometry" &&
                        ! col. ColumnName. StartsWith("Envelope_"))
                        fdt. Columns. Add(col. ColumnName, col. DataType, col. Expression);
                foreach (DataRow dr in ds2. Tables[0]. Rows)
                {
                    FeatureDataRow fdr = fdt. NewRow();
                    foreach (DataColumn col in ds2. Tables[0]. Columns)
                        if (col. ColumnName ! = GeometryColumn && col. ColumnName ! = "sharpmap_tempgeometry" &&
                            ! col. ColumnName. StartsWith("Envelope_"))
                            fdr[col. ColumnName] = dr[col];
                    if (dr["sharpmap_tempgeometry"] ! = DBNull. Value)
                        fdr. Geometry = GeometryFromWKB. Parse((byte[]) dr["sharpmap_
```

```
                    tempgeometry"]);
                                fdt.AddRow(fdr);
            }
                    ds.Tables.Add(fdt);
        }
    }
}
```

7.5.3.12 Dispose()

实现 Dispose 设计模式的资源的回收代码,代码如下:

```
public void Dispose()
{
    Dispose(true);
    GC.SuppressFinalize(this);
}
```

7.5.3.13 Dispose(bool disposing)

实现 Dispose 模式的方法,代码如下:

```
internal void Dispose(bool disposing)
{
    if(! disposed)
    {
        if(disposing)
        {
        }
        disposed = true;
    }
}
```

7.5.3.14 ~MsSql()

析构函数,保证资源的及时释放和垃圾收集,代码如下:

```
~MsSql()
{
    Dispose();
}
```

7.5.3.15 GetBoxClause(BoundingBox bbox)

根据给定的外包框矩形,生成在此外包框内的限制语句,代码如下:

```
private string GetBoxClause(BoundingBox bbox)
{
    return String.Format(Map.NumberFormatEnUs, "(Envelope_MinX < {0} AND Envelope_
```

MaxX > {1} AND Envelope_MinY < {2} AND Envelope_MaxY > {3})", bbox.Max.X, bbox.Min.X, bbox.Max.Y, bbox.Min.Y);
}

7.5.3.16 CreateDataTable (IProvider datasource, string tablename, string connstr)

在 microsoft Sql Server 中创建数据表，并从指定的数据源中拷贝要素数据到这个新生成的数据表中。首先调用数据源的 Open 方法打开数据源，调用数据源的 GetFeature 方法，设置参数为 0，获取第 0 个要素数据行，保存于 FeatureDataRow 类型的变量 geom 中，获取其列的集合保存于 columns 中，代码如下：

```
public static int CreateDataTable(IProvider datasource, string tablename, string connstr)
{
    datasource.Open();
    FeatureDataRow geom = datasource.GetFeature(0);
    DataColumnCollection columns = geom.Table.Columns;
```

设置整形变量 counter 初始值为 -1，新建一个 Sql 命令 command，设置连接属性值为 conn，打开连接，设置命令文本为参数 tablename 指定的表，然后调用其 ExecuteNonQuery 方法执行命令：

```
    int counter = -1;
    using (SqlConnection conn = new SqlConnection(connstr))
    {
        SqlCommand command = new SqlCommand();
        command.Connection = conn;
        conn.Open();
        try
        {
            command.CommandText = "DROP TABLE \"" + tablename + "\";";
            command.ExecuteNonQuery();
        }
        catch
        {
        }
        //Create new table for storing the datasource
```

新建一个字符串对象 sql，用于保存创建参数 tablename 指定的表名的数据表的 sql 命令，可以看出，新建的表具有 oid、Geometry、Envelope_MinX、Envelope_MinY、Envelope_MaxX、Envelope_MaxY 六个基本属性列，然后遍历 columns 集合，加入 columns 集合中的每个属性命令，最后重新设置 command 命令文本属性为新的 sql，调用命令对象的 ExecuteNonQuery 方法执行创建工作，并将 counter 自加 1：

```
        string sql = "CREATE TABLE " + tablename + " (oid INTEGER IDENTITY PRIMARY KEY, WKB_Geometry Image, " +
            "Envelope_MinX real, Envelope_MinY real, Envelope_MaxX real, Envelope_MaxY re-
```

```
                al";
    foreach (DataColumn col in columns)
        if (col. DataType ! = typeof (String))
            sql + = "," + col. ColumnName + " " + Type2SqlType(col. DataType). ToString();
        else
            sql + = "," + col. ColumnName + " VARCHAR(256)";
    command. CommandText = sql + ");";
    command. ExecuteNonQuery();
    counter + +;
```

调用数据源的 GetObjectIDsInView 方法获取整个数据源范围内（datasource. GetExtents()）的要素的 ID 编号的集合，然后遍历这个编号的集合，调用数据源的 GetFeature 方法获取对应的要素数据行，若 counter 值为 0，生成将要素插入数据表的语句：

```
Collection < uint > indexes = datasource. GetObjectIDsInView(datasource. GetExtents());
foreach (uint idx in indexes)
{
    //Get feature from shapefile
    FeatureDataRow feature = datasource. GetFeature(idx);
    if (counter = = 0)
    {
        //Create insert script
        string strSQL = " (";
        foreach (DataColumn col in feature. Table. Columns)
            strSQL + = "@" + col. ColumnName + ",";
        strSQL + = "@ WKB_Geometry,@ Envelope_MinX,@ Envelope_MinY, " +
                  "@ Envelope_MaxX,@ Envelope_MaxY)";
        strSQL = "INSERT INTO " + tablename + strSQL. Replace("@","") + " VALUES" + strSQL;
        command. CommandText = strSQL;
        command. Parameters. Clear();
        foreach (DataColumn col in feature. Table. Columns)
            command. Parameters. Add("@" + col. ColumnName, Type2SqlType(col. DataType));
        //Add geometry parameters
        command. Parameters. Add("@ WKB_Geometry", SqlDbType. VarBinary);
        command. Parameters. Add("@ Envelope_MinX", SqlDbType. Real);
        command. Parameters. Add("@ Envelope_MinY", SqlDbType. Real);
        command. Parameters. Add("@ Envelope_MaxX", SqlDbType. Real);
        command. Parameters. Add("@ Envelope_MaxY", SqlDbType. Real);
    }
```

设置要插入的值，若要素的 Geometry 属性不为空，设置@ Envelope_MinX、@ Envelope_MinY、@ Envelope_MaxX、@ Envelope_MaxY 的值分别为要素外包框矩形的左、上、右、下值；若为空，则将其都设置为 DBNull. Value，然后调用命令对象的 ExecuteNonQuery 方法执行命令，将 counter 自加 1：

```
    foreach (DataColumn col in feature.Table.Columns)
    command.Parameters["@" + col.ColumnName].Value = feature[col];
    if (feature.Geometry != null)
    {
        command.Parameters["@WKB_Geometry"].Value = feature.Geometry.AsBinary();
        //Add the geometry as Well-Known Binary
        BoundingBox box = feature.Geometry.GetBoundingBox();
        command.Parameters["@Envelope_MinX"].Value = box.Left;
        command.Parameters["@Envelope_MinY"].Value = box.Bottom;
        command.Parameters["@Envelope_MaxX"].Value = box.Right;
        command.Parameters["@Envelope_MaxY"].Value = box.Top;
    }
    else
    {
        command.Parameters["@WKB_Geometry"].Value = DBNull.Value;
        command.Parameters["@Envelope_MinX"].Value = DBNull.Value;
        command.Parameters["@Envelope_MinY"].Value = DBNull.Value;
        command.Parameters["@Envelope_MaxX"].Value = DBNull.Value;
        command.Parameters["@Envelope_MaxY"].Value = DBNull.Value;
    }
    //Insert row
    command.ExecuteNonQuery();
        counter++;
}
```

每插入一个对象,就在数据表的 Envelope_MinX、Envelope_MinY、Envelope_MaxX、Envelope_MaxY 四个列上创建索引,最后关闭数据源,返回 counter 值:

```
        command.Parameters.Clear();
        command.CommandText = "CREATE INDEX [IDX_Envelope_MinX] ON " + tablename + " (Envelope_MinX)";
        command.ExecuteNonQuery();
        command.CommandText = "CREATE INDEX [IDX_Envelope_MinY] ON " + tablename + " (Envelope_MinY)";
        command.ExecuteNonQuery();
        command.CommandText = "CREATE INDEX [IDX_Envelope_MaxX] ON " + tablename + " (Envelope_MaxX)";
        command.ExecuteNonQuery();
        command.CommandText = "CREATE INDEX [IDX_Envelope_MaxY] ON " + tablename + " (Envelope_MaxY)";
        command.ExecuteNonQuery();
        conn.Close();
    }
    datasource.Close();
```

```
return counter;
}
```

7.5.3.17 Type2SqlType（Type t）

将 .NET 的类型转换为 Sql 数据库的类型，代码如下：

```
private static SqlDbType Type2SqlType( Type t)
{
    switch (t. ToString( ))
    {
        case "System. Boolean":
            return SqlDbType. Bit;
        case "System. Single":
            return SqlDbType. Real;
        case "System. Double":
            return SqlDbType. Float;
        case "System. Int16":
            return SqlDbType. SmallInt;
        case "System. Int32":
            return SqlDbType. Int;
        case "System. Int64":
            return SqlDbType. BigInt;
        case "System. DateTime":
            return SqlDbType. DateTime;
        case "System. Byte[ ]":
            return SqlDbType. Image;
        case "System. String":
            return SqlDbType. VarChar;
        default:
            throw ( new NotSupportedException("Unsupported datatype '" + t. Name + "' found in datasource"));
    }
}
```

7.6 其他 Provider 类

GeometryFeatureProvider、GeometryProvider、MsSqlSpatial、OleDbPoint、SqlServer2008、WFS 等类原理相似，由于篇幅关系，请读者自行研究。

7.7 FeatureDataSet 类

继承自 .NET 的 System. Data. DataSet，要素数据集，可包含多个要素表 FeatureDataTable 对象，并且加上了 Serializable 特性，说明此类是可以序列化的。

7.7.1 FeatureDataSet 类的属性

Tables 是获取和设置数据表的集合，这是一个 FeatureTableCollection（就是 List < FeatureDataTable >）类型的对象，用于保存数据表的集合，代码如下：

```
private FeatureTableCollection _FeatureTables;
public new FeatureTableCollection Tables
{
    get { return _FeatureTables; }
}
```

7.7.2 FeatureDataSet 类的方法

7.7.2.1 FeatureDataSet（ ）

构造函数。这里调用 InitClass（ ）方法初始化表集合，为 Relations 附加了事件处理函数 SchemaChanged，代码如下：

```
public FeatureDataSet()
{
    InitClass();
    CollectionChangeEventHandler schemaChangedHandler = new CollectionChangeEventHandler
(SchemaChanged);
    Relations.CollectionChanged + = schemaChangedHandler;
    InitClass();
}
```

7.7.2.2 FeatureDataSet（SerializationInfo info，StreamingContext context）

构造函数。先读取 XML 架构信息，若架构信息不为空，则将新生成的数据集 DataSet 合并到当前的 DataSet，若为空则调用 InitClass（ ）方法初始化类信息；最后调用 GetSerializationData（ ）方法从二进制或 XML 流反序列化表数据，填充当前的数据集，代码如下：

```
protected FeatureDataSet(SerializationInfo info, StreamingContext context)
{
    string strSchema = ((string)(info.GetValue("XmlSchema", typeof(string))));
    if ((strSchema ! = null))
    {
        DataSet ds = new DataSet();
        ds.ReadXmlSchema(new XmlTextReader(new StringReader(strSchema)));
        if (((ds.Tables["FeatureTable"] ! = null))
        {
            Tables.Add(new FeatureDataTable(ds.Tables["FeatureTable"]));
        }
```

```
            DataSetName = ds.DataSetName;
            Prefix = ds.Prefix;
            Namespace = ds.Namespace;
            Locale = ds.Locale;
            CaseSensitive = ds.CaseSensitive;
            EnforceConstraints = ds.EnforceConstraints;
            Merge(ds, false, MissingSchemaAction.Add);
        }
    else
        {
            InitClass();
        }
        GetSerializationData(info, context);
        CollectionChangeEventHandler schemaChangedHandler = new CollectionChangeEventHandler
(SchemaChanged);
        Relations.CollectionChanged += schemaChangedHandler;
    }
```

7.7.2.3 Clone()

要素数据集拷贝。这里仅仅实现了数据的浅拷贝，只拷贝了要素数据集的架构：要素集表的架构、表之间的关系、约束关系，代码如下：

```
public new FeatureDataSet Clone()
{
    FeatureDataSet cln = ((FeatureDataSet)(base.Clone()));
    return cln;
}
```

7.7.2.4 ShouldSerializeTables()

表示数据表能否被存储（序列化）。当前版本中，设定为否，代码如下：

```
protected override bool ShouldSerializeTables()
{
    return false;
}
```

7.7.2.5 ShouldSerializeRelations()

表示表之间的关系能否被持久化（序列化）。当前版本中，设定为否，代码如下：

```
protected override bool ShouldSerializeRelations()
{
    return false;
}
```

7.7.2.6 ReadXmlSerializable(XmlReader reader)

读取经过 Xml 格式序列化后的数据，将读取的数据合并到当前的要素集合中，代码

如下：

```csharp
protected override void ReadXmlSerializable(XmlReader reader)
{
    Reset();
    DataSet ds = new DataSet();
    ds.ReadXml(reader);
    DataSetName = ds.DataSetName;
    Prefix = ds.Prefix;
    Namespace = ds.Namespace;
    Locale = ds.Locale;
    CaseSensitive = ds.CaseSensitive;
    EnforceConstraints = ds.EnforceConstraints;
    Merge(ds, false, MissingSchemaAction.Add);
}
```

7.7.2.7 GetSchemaSerializable()

得到数据表的架构信息，调用 DataSet 的 WriteXmlSchema() 方法将 DataTable 的架构写入 XML 文档。架构包括表、关系和约束定义，最后调用 XmlSchema.Read() 方法读取文件架构，代码如下：

```csharp
protected override XmlSchema GetSchemaSerializable()
{
    MemoryStream stream = new MemoryStream();
    WriteXmlSchema(new XmlTextWriter(stream, null));
    stream.Position = 0;
    return XmlSchema.Read(new XmlTextReader(stream), null);
}
```

7.7.2.8 InitClass()

初始化类。在这个方法中，初始化了一个新的要素表集合；并初始化父类相关成员，代码如下：

```csharp
private void InitClass()
{
    _FeatureTables = new FeatureTableCollection();
    //this.DataSetName = "FeatureDataSet";
    Prefix = "";
    Namespace = "http://tempuri.org/FeatureDataSet.xsd";
    Locale = new CultureInfo("en-US");
    CaseSensitive = false;
    EnforceConstraints = true;
}
```

7.7.2.9 ShouldSerializeFeatureTable()

得到一个表示数据表能否被存储（序列化）的值，这里总是返回 false。代码如下：

```
private bool ShouldSerializeFeatureTable()
{
    return false;
}
```

7.7.2.10 SchemaChanged(object sender, CollectionChangeEventArgs e)

架构变化时事件处理函数。当前版本中实际上什么都没做，代码如下：

```
private void SchemaChanged(object sender, CollectionChangeEventArgs e)
{
    if ((e.Action == CollectionChangeAction.Remove))
    {
    }
}
```

7.8 FeatureDataTable 类

继承自 DataTable，表示空间要素集，并且标记 Serializable 特性，为可序列化对象。

7.8.1 FeatureDataTable 类的属性

7.8.1.1 Count

获取要素数据行的个数，代码如下：

```
[Browsable(false)]
public int Count
{
    get { return base.Rows.Count; }
}
```

7.8.1.2 this[int index]

通过索引值获取数据行，代码如下：

```
public FeatureDataRow this[int index]
{
    get { return (FeatureDataRow)base.Rows[index]; }
}
```

7.8.2 FeatureDataTable 类的方法

7.8.2.1 AddRow(FeatureDataRow row)

添加数据行，代码如下：

```
public void AddRow(FeatureDataRow row)
{
```

```
            base.Rows.Add(row);
    }
```

7.8.2.2 Clone()

数据表的拷贝,包括数据表的架构、限制条件等,但不包括实际的数据,代码如下:

```
    public new FeatureDataTable Clone()
    {
        FeatureDataTable cln = ((FeatureDataTable)(base.Clone()));
        cln.InitVars();
        return cln;
    }
```

7.8.2.3 CreateInstance()

创建数据表的一个新的实例,与本身的数据表并无关系,代码如下:

```
    protected override DataTable CreateInstance()
    {
        return new FeatureDataTable();
    }
```

7.8.2.4 InitVars()

并无实际的作用,代码如下:

```
    internal void InitVars()
    {
    }
```

7.8.2.5 InitClass()

这个方法也没有任何实际的意义,代码如下:

```
    private void InitClass()
    {
    }
```

7.8.2.6 NewRow()

产生一个新的数据行。这里实际上调用 System.Data.DataTable 的 NewRow 方法产生一个新的数据行,并强制转换为要素数据行 FeatureDataRow,代码如下:

```
    public new FeatureDataRow NewRow()
    {
        return (FeatureDataRow)base.NewRow();
    }
```

7.8.2.7 NewRowFromBuilder(DataRowBuilder builder)

通过 System.Data.DataRowBuilder 创建一个新的数据行,代码如下:

```
    protected override DataRow NewRowFromBuilder(DataRowBuilder builder)
```

```
        return new FeatureDataRow(builder);
}
```

7.8.2.8 GetRowType()

获取数据行的类型。这里调用 typeof 操作符，代码如下：

```
protected override Type GetRowType()
{
        return typeof(FeatureDataRow);
}
```

7.8.3 FeatureDataTable 类的事件

这里首先声明了一个 FeatureDataRowChangeEventHander 委托，用于事件处理操作。代码如下：

```
public delegate void FeatureDataRowChangeEventHandler(object sender, FeatureDataRowChangeEventArgs e);
```

7.8.3.1 FeatureDataRowChanged

当数据行变化后会引发这个事件，代码如下：

```
public event FeatureDataRowChangeEventHandler FeatureDataRowChanged;
```

7.8.3.2 FeatureDataRowChanging

当数据行正在变化时会引发这个事件，代码如下：

```
public event FeatureDataRowChangeEventHandler FeatureDataRowChanging;
```

7.8.3.3 FeatureDataRowDeleted

当数据行被删除后会引发这个事件，代码如下：

```
public event FeatureDataRowChangeEventHandler FeatureDataRowDeleted;
```

7.8.3.4 FeatureDataRowDeleting

当数据行正在被删除时，会引发这个事件，代码如下：

```
public event FeatureDataRowChangeEventHandler FeatureDataRowDeleting;
```

7.9 FeatureDataRow 类

要素对象，代表要素集中的一个对象，派生自 DataRow 类。

7.9.1 FeatureDataRow 类的属性

Geometry 为当前要素的几何对象，代码如下：

```
        private Geometry _Geometry;
        public Geometry Geometry
        {
            get { return _Geometry; }
            set { _Geometry = value; }
        }
```

7.9.2 FeatureDataRow 类的方法

7.9.2.1 FeatureDataRow (DataRowBuilder rb)

通过 DataRowBuilder 构建一个新的数据行实例。这里直接调用父类 DataRow 的构造函数进行创建工作，代码如下：

```
        internal FeatureDataRow(DataRowBuilder rb) : base(rb)
        {
        }
```

7.9.2.2 IsFeatureGeometryNull()

获取一个值，该值指示要素的几何对象是否为空，代码如下：

```
        public bool IsFeatureGeometryNull()
        {
            return Geometry == null;
        }
```

7.9.2.3 SetFeatureGeometryNull()

将要素的几个对象设置为空，代码如下：

```
        public void SetFeatureGeometryNull()
        {
            Geometry = null;
        }
```

复习思考题

7-1 思考 SharpMap 中空间数据连接池的意义。

7-2 数据提供接口 IProvider 主要功能是什么，这么设计有什么好处？

7-3 试利用 FeatureDataSet、FeatureDataTable、FeatureDataRow 实现空间数据查询功能。

第8章 几何对象

空间数据包括几何信息、属性信息，其中几何信息是空间数据区别于一般数据的主要内容。SharpMap 中几何对象（Geometry）位于 SharpMap.Geometries 名称空间下。

Geometry 对象是对现实世界中的各种事物抽象出来的几何描述，从概念上一般分为三类，就是点、线、面三种类型。应该注意还有一个词汇 Feature，一般译为要素。要素是对现实世界中的一个对象的完整描述，除了用几何信息，还应该有属性信息。例如，一个城市，可以用一个多边形描述其几何信息，再加上城市的名称、人口数、面积等其他属性，就构成了 Feature。简单的说，就是 Geometry 加上相关的属性信息构成 Feature。

从抽象的角度来看，系统的几何类型是越少越好，但是从功能和兼容其他数据的角度来看，系统的几何类型是越丰富越好，这是一对矛盾。

空间分析中主要使用的就是 Geometry 对象，这是很容易理解的，很多分析都是直接在 Geometry 类的基础上做的，如点、线、面的缓冲区就可以如此。空间关系的判定也需要使用 Geometry 对象，比方说要判断对象是否相交时，如果没有构建两个 Geometry，就没办法判断这两个几何体是否相交。Geometries 模块设计遵循 OGC OpenGIS Simple Feature Specification 规范。SharpMap 中几何对象（Geometry）类关系图如图 8-1 所示。

IGeometry 接口中定义了所有几何对象必须实现的功能，Geometry 抽象类则实现了所有几何对象通用的属性和方法。所有几何形状都继承 Geometry，如 Point、Curve、Surface 等。

IGeometryCollection 从 IGeometry 接口派生，用于表达复合几何对象，即由多个简单几何对象组合而成，如 MultiCurve、MultiPoint、MultiSurface 等。

BoundingBox 对象表达的是一个矩形，用于代表几何对象的外接矩形，它并不继承 Geometry 类。

枚举类型 GeometryType2 包含了所有几何对象。

SpatialRelations 类中则定义了常用的空间关系判别功能，如 Contains（包含）、Intersects（相交）、Overlaps（重叠）等。

由于篇幅关系，本书只讲述 Geometry、Point、MultiPoint、Curve、Polygon、BoundingBox、SpatialRelations 类。其他内容原理与本书讲述内容相似，请读者参考源代码。

8.1 几何对象抽象基类 Geometry

8.1.1 Geometry 的属性

8.1.1.1 SpatialReference

SpatialReference 为几何对象的空间参考，它使用了开源投影类库 Proj4 的 ICoordinate-

8.1 几何对象抽象基类 Geometry

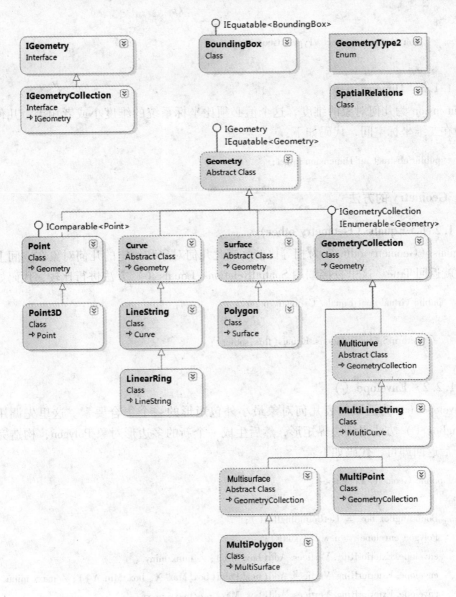

图 8-1 SharpMap 中几何对象 (Geometry) 类图

System 接口，代码如下：

```
private ICoordinateSystem _SpatialReference;
public ICoordinateSystem SpatialReference
{
    get { return _SpatialReference; }
    set { _SpatialReference = value; }
}
```

8.1.1.2 GeometryType

GeometryType 为几何对象的类型。这里使用的是 GeometryType2 枚举值，代码如下：

```
public virtual GeometryType2 GeometryType
```

```
            get { return GeometryType2.Geometry; }
        }
```

8.1.1.3 Dimension

Dimension 为几何对象的维度,这个值必须比坐标系统的维度小或者相等,几何对象被限制在二维坐标空间,代码如下:

```
        public abstract int Dimension { get; }
```

8.1.2 Geometry 的方法

8.1.2.1 Equals (Geometry other)

Equals (Geometry other) 为相等比较,当此几何对象和给定的几何对象在空间上完全相等时就返回 True。这里直接调用 SpatialRelations.Equals() 方法进行比较,代码如下:

```
        public virtual bool Equals(Geometry other)
        {
            return SpatialRelations.Equals(this, other);
        }
```

8.1.2.2 Envelope()

Envelope() 为返回代表几何对象最小外包矩形的一个集合要素。这里先调用 GetBoundingBox() 方法得到边界矩形,然后生成一个新的多边形对象 Polygon,构造完这个 Polygon 后返回即可,代码如下:

```
        public Geometry Envelope()
        {
            BoundingBox box = GetBoundingBox();
            Polygon envelope = new Polygon();
            envelope.ExteriorRing.Vertices.Add(box.Min); //minx miny
            envelope.ExteriorRing.Vertices.Add(new Point(box.Max.X, box.Min.Y)); //maxx minu
            envelope.ExteriorRing.Vertices.Add(box.Max); //maxx maxy
            envelope.ExteriorRing.Vertices.Add(new Point(box.Min.X, box.Max.Y)); //minx maxy
            envelope.ExteriorRing.Vertices.Add(envelope.ExteriorRing.StartPoint); //close ring
            return envelope;
        }
```

8.1.2.3 GetBoundingBox()

GetBoundingBox() 代表几何对象最小外包矩形的一个边界矩形,代码如下:

```
        public abstract BoundingBox GetBoundingBox();
```

8.1.2.4 AsText()

AsText() 代表返回当前几何对象的文本表示(二维点对象的文本表示就是点坐标的值),代码如下:

8.1 几何对象抽象基类 Geometry

```
public string AsText()
{
    return GeometryToWKT.Write(this);
}
```

8.1.2.5 AsBinary()

AsBinary()表示将当前几何对象用小位序编码写入字节数组,代码如下:

```
public byte[] AsBinary()
{
    return GeometryToWKB.Write(this);
}
```

8.1.2.6 ToString()

ToString()代表返回当前几何对象的文本表示,它在内部实际上调用 AsText()方法,代码如下:

```
public override string ToString()
{
    return AsText();
}
```

8.1.2.7 IsEmpty()

当前几何对象是一个空对象时就为 True,表示在坐标空间中当前几何对象是一个空点集,代码如下:

```
public abstract bool IsEmpty();
```

8.1.2.8 IsSimple()

是否是简单的几何对象,当几何对象没有包含异常点,比如自相交或相邻时就为 True,代码如下:

```
public abstract bool IsSimple();
```

8.1.2.9 Boundary()

Boundary()表示返回几何对象的边界对象,这个边界对象是自封闭的,代码如下:

```
public abstract Geometry Boundary();
```

8.1.2.10 GeomFromText(string WKT)

GeomFromText(string WKT)表示将字符串转换为几何对象。这里的转换工作是调用 SharpMap.Converters.WellKnownText 中的 Parse()方法,代码如下:

```
public static Geometry GeomFromText(string WKT)
{
    return GeometryFromWKT.Parse(WKT);
}
```

8.1.2.11 GeomFromWKB (byte [ ] WKB)

GeomFromWKB (byte [ ] WKB) 表示从字节数组中创建一个几何对象,代码如下:

```
public static Geometry GeomFromWKB(byte[ ] WKB)
{
    return GeometryFromWKB.Parse(WKB);
}
```

8.1.2.12 Disjoint (Geometry geom)

当前几何对象与给定的几何对象不相交时就为 True。这里调用 SpatialRelations 类的 Disjoint 方法来进行判断,代码如下:

```
public virtual bool Disjoint(Geometry geom)
{
    return SpatialRelations.Disjoint(this, geom);
}
```

8.1.2.13 Intersects (Geometry geom)

当前几何对象与给定的几何对象相交时就为 True。这里调用 SpatialRelations 类的 Intersects 方法来进行判断,代码如下:

```
public virtual bool Intersects(Geometry geom)
{
    return SpatialRelations.Intersects(this, geom);
}
```

8.1.2.14 Touches (Geometry geom)

当前几何对象与给定的几何对象接触时就为 True。这里调用 SpatialRelations 类的 Touches 方法来进行判断,代码如下:

```
public virtual bool Touches(Geometry geom)
{
    return SpatialRelations.Touches(this, geom);
}
```

8.1.2.15 Crosses (Geometry geom)

当前几何对象与给定的几何对象交叉时就为 True。这里调用 SpatialRelations 类的 Crosses 方法来进行判断,代码如下:

```
public virtual bool Crosses(Geometry geom)
{
    return SpatialRelations.Crosses(this, geom);
}
```

8.1.2.16 Within (Geometry geom)

当前几何对象在给定的几何对象内部时就为 True。这里调用 SpatialRelations 类的

Within 方法来进行判断，代码如下：

```
public virtual bool Within(Geometry geom)
{
    return SpatialRelations.Within(this, geom);
}
```

8.1.2.17　Contains（Geometry geom）

当前几何对象包含给定的几何对象时就为 True。这里调用 SpatialRelations 类的 Contains 方法来进行判断，代码如下：

```
public virtual bool Contains(Geometry geom)
{
    return SpatialRelations.Contains(this, geom);
}
```

8.1.2.18　Overlaps（Geometry geom）

当前几何对象与给定的几何对象相叠时就为 True。这里调用 SpatialRelations 类的 Overlaps 方法来进行判断，代码如下：

```
public virtual bool Overlaps(Geometry geom)
{
    return SpatialRelations.Overlaps(this, geom);
}
```

8.1.2.19　Relate（Geometry other, string intersectionPattern）

Relate（Geometry other, string intersectionPattern） 表示空间关联，当前几何对象与给定的几何对象有空间关系时就为 True。目前并未实现该功能，代码如下：

```
public bool Relate(Geometry other, string intersectionPattern)
{
    throw new NotImplementedException();
}
```

8.1.2.20　Distance（Geometry geom）

Distance（Geometry geom） 表示返回两个几何对象之间最小距离，代码如下：

```
public abstract double Distance(Geometry geom);
```

8.1.2.21　Buffer（double d）

Buffer（double d） 表示返回几何对象指定距离的缓冲区对象，代码如下：

```
public abstract Geometry Buffer(double d);
```

8.1.2.22　ConvexHull（）

ConvexHull（） 表示返回这个几何对象的"凸壳"，代码如下：

```
public abstract Geometry ConvexHull();
```

8.1.2.23 Intersection（Geometry geom）

Intersection（Geometry geom）表示返回当前几何对象和指定的几何对象（传入的参数）相交部分的几何对象，代码如下：

```
public abstract Geometry Intersection(Geometry geom);
```

8.1.2.24 Union（Geometry geom）

Union（Geometry geom）表示返回当前几何对象和指定的几何对象（传入的参数）并集的几何对象，代码如下：

```
public abstract Geometry Union(Geometry geom);
```

8.1.2.25 Difference（Geometry geom）

Difference（Geometry geom）表示返回当前几何对象和指定的几何对象（传入的参数）不同的几何对象，代码如下：

```
public abstract Geometry Difference(Geometry geom);
```

8.1.2.26 SymDifference（Geometry geom）

SymDifference（Geometry geom）表示返回当前几何对象和指定的几何对象（传入的参数）对称不同的几何对象，代码如下：

```
public abstract Geometry SymDifference(Geometry geom);
```

8.1.2.27 Clone（）

Clone（）表示当前几何对象的拷贝。这里并未实现，需要在子类中重写，代码如下：

```
public Geometry Clone()
{
    throw (new ApplicationException("Clone() has not been implemented on derived datatype"));
}
```

8.1.2.28 Equals（object obj）

Equals（object obj）为当前对象和指定的对象是否相等的比较，代码如下：

```
public override bool Equals(object obj)
{
    if (obj == null)
        return false;
    else
    {
        Geometry g = obj as Geometry;
        if (g == null)
            return false;
        else
            return Equals(g);
    }
}
```

8.1.2.29 GetHashCode（ ）

GetHashCode（ ）表示返回当前对象二进制数据的哈希值，代码如下：

```
public override int GetHashCode()
{
    return AsBinary().GetHashCode();
}
```

8.2 点对象 Point

8.2.1 Point 的数据成员

_IsEmpty 表示点对象是否为空，代码如下：

```
private bool _IsEmpty = false;
```

8.2.2 Point 的属性

8.2.2.1 GeometryType

GeometryType 为集合类型。这里使用的是 GeometryType2.Point 枚举值，代表当前对象是一个点，代码如下：

```
public override GeometryType2 GeometryType
{
    get
    {
        return GeometryType2.Point;
    }
}
```

8.2.2.2 Dimension

Dimension 表示几何对象的维度。由于是点对象，这里返回 0，表示是一个 0 维的对象，代码如下：

```
public override int Dimension
{
    get { return 0; }
}
```

8.2.2.3 NumOrdinates

NumOrdinates 为点的分布序列的数值。这里返回整形数据 2，代码如下：

```
public virtual int NumOrdinates
{
```

```
        get { return 2; }
}
```

8.2.2.4 SetIsEmpty

SetIsEmpty 表示把当前的点对象设置为空，代码如下：

```
protected bool SetIsEmpty
{
    set { _IsEmpty = value; }
}
```

8.2.2.5 X

X 表示获取和设置点的 X 坐标值，代码如下：

```
private double _X;
public double X
{
    get
    {
        if (!_IsEmpty) return _X;
        else throw new ApplicationException("Point is empty");
    }
    set
    {
        _X = value;
        _IsEmpty = false;
    }
}
```

8.2.2.6 Y

Y 表示获取和设置点的 Y 坐标值，代码如下：

```
private double _Y;
public double Y
{
    get
    {
        if (!_IsEmpty)
            return _Y;
        else throw new ApplicationException("Point is empty");
    }
    set
    {
        _Y = value;
        _IsEmpty = false;
```

8.2 点对象 Point

```
    }
}
```

8.2.2.7 this [uint index]

this [uint index] 表示通过索引得到坐标值。这里索引的最大值为1，否则会跑出异常。0代表X值，1代表Y值，代码如下：

```
public virtual double this[uint index]
{
    get
    {
        if (_IsEmpty) throw new ApplicationException("Point is empty");
        else if (index == 0) return X;
        else if(index == 1) return Y;
        else throw (new Exception("Point index out of bounds"));
    }
    set
    {
        if (index == 0) X = value;
        else if (index == 1) Y = value;
        else throw (new Exception("Point index out of bounds"));
        _IsEmpty = false;
    }
}
```

8.2.3 Point 的方法

8.2.3.1 Point (double x, double y)

此为构造函数，通过指定的X、Y值来初始化一个点对象，代码如下：

```
public Point(double x, double y)
{
    _X = x;
    _Y = y;
}
```

8.2.3.2 Point ()

这是默认的构造函数。这里默认生成一个X、Y坐标值都为0的点对象，代码如下：

```
public Point() : this(0, 0)
{
    _IsEmpty = true;
}
```

8.2.3.3 Point (double [] point)

Point (double [] point) 表示用一个Double类型的数组实例化一个点对象，当然，这个点数组的长度必须大于或等于2，否则会跑出异常，代码如下：

```
public Point(double[] point)
{
    if (point.Length < 2)
        throw new Exception("Only 2 dimensions are supported for points");
    _X = point[0];
    _Y = point[1];
}
```

8.2.3.4 CompareTo（Point other）

CompareTo（Point other）表示当前点对象与指定的点对象比较，根据比较的结果返回不同的值，代码如下：

```
public virtual int CompareTo(Point other)
{
    if (X < other.X || X == other.X && Y < other.Y)
        return -1;
    else if (X > other.X || X == other.X && Y > other.Y)
        return 1;
    else // (this.X == other.X && this.Y == other.Y)
        return 0;
}
```

8.2.3.5 Equals（Point p）

Equals（Point p）表示当前点对象与指定的点对象是否相等的比较。这里比较的就是两个点对象对应的 X、Y 值是否相等，代码如下：

```
public virtual bool Equals(Point p)
{
    return p != null && p.X == _X && p.Y == _Y && _IsEmpty == p.IsEmpty();
}
```

8.2.3.6 GetHashCode（）

GetHashCode（）表示得到当前点对象的哈希值，具体的哈希值的计算方法如下面代码，非常简单，这里就不再赘述了。

```
public override int GetHashCode()
{
    return _X.GetHashCode() ^ _Y.GetHashCode() ^ _IsEmpty.GetHashCode();
}
```

8.2.3.7 IsEmpty（）

IsEmpty（）表示当前点对象是否为空，代码如下：

```
public override bool IsEmpty()
```

8.2 点对象 Point

```
        return _IsEmpty;
    }
```

8.2.3.8 IsSimple ( )

IsSimple ( ) 表示当前点对象是否是简单对象。由于点对象不可能出现类似自相交的情况，它是最简单的空间对象。这里直接返回 True，代码如下：

```
public override bool IsSimple( )
{
    return true;
}
```

8.2.3.9 Boundary ( )

Boundary ( ) 表示点对象的边界对象。这里返回空，因为点对象是没有边界的，代码如下：

```
public override Geometry Boundary( )
{
    return null;
}
```

8.2.3.10 Distance (Geometry geom)

Distance (Geometry geom) 表示返回当前点对象与指定的几何对象之间的距离。目前只实现了和点对象之间距离的计算即两点之间距离，对于其他对象这里抛出异常，代码如下：

```
public override double Distance( Geometry geom)
{
    if ( geom.GetType( ) == typeof ( Point) )
    {
        Point p = geom as Point;
        return Math.Sqrt( Math.Pow( X - p.X, 2) + Math.Pow( Y - p.Y, 2) );
    }
    else
        throw new Exception( "The method or operation is not implemented for this geometry type." );
}
```

8.2.3.11 Distance (BoundingBox box)

Distance (BoundingBox box) 表示返回当前点对象与指定的边界矩形之间的距离。这里直接调用 BoundingBox 的 Distance ( ) 方法进行计算，代码如下：

```
public double Distance( BoundingBox box)
{
    return box.Distance( this) ;
}
```

8.2.3.12 Buffer (double d)

Buffer (double d) 表示点的缓冲几何对象。目前并未实现，代码如下：

```
public override Geometry Buffer(double d)
{
    throw new NotImplementedException();
}
```

8.2.3.13 ConvexHull()

ConvexHull() 表示返回点对象的凸壳。由于这个操作对于点对象并无实际的意义，这里抛出异常，代码如下：

```
public override Geometry ConvexHull()
{
    throw new NotImplementedException();
}
```

8.2.3.14 Intersection (Geometry geom)

Intersection (Geometry geom) 表示返回点对象相交的几何对象。这里并未实现，代码如下：

```
public override Geometry Intersection(Geometry geom)
{
    throw new NotImplementedException();
}
```

8.2.3.15 Union (Geometry geom)

Union (Geometry geom) 表示点对象和指定几何对象的并集。这里并未实现，代码如下：

```
public override Geometry Union(Geometry geom)
{
    throw new NotImplementedException();
}
```

8.2.3.16 Difference (Geometry geom)

Difference (Geometry geom) 表示点对象和指定的几何对象的不同。这里并未实现，代码如下：

```
public override Geometry Difference(Geometry geom)
{
    throw new NotImplementedException();
}
```

8.2.3.17 SymDifference (Geometry geom)

SymDifference (Geometry geom) 表示点对象与指定的几何对象对称不同。这里并未实现，代码如下：

```
public override Geometry SymDifference(Geometry geom)
{
    throw new NotImplementedException();
}
```

8.2.3.18 GetBoundingBox()

GetBoundingBox() 表示点对象的边界框。这里生成一个新的边界矩形框后返回，这个边界矩形实际上就是一个点，代码如下：

```
public override BoundingBox GetBoundingBox()
{
    return new BoundingBox(X, Y, X, Y);
}
```

8.2.3.19 Touchcs (BoundingBox box)

Touches (BoundingBox box) 表示检查当前点对象是否与指定的边界矩形框接触。这里调用 BoundingBox 的 Touchs() 方法进行接触检查，代码如下：

```
public bool Touches(BoundingBox box)
{
    return box.Touches(this);
}
```

8.2.3.20 Touches (Geometry geom)

Touches (Geometry geom) 表示检查当前点对象是否与指定的几何对象接触。这里并未实现，代码如下：

```
public override bool Touches(Geometry geom)
{
    if (geom is Point && Equals(geom)) return true;
    throw new NotImplementedException("Touches not implemented for this feature type");
}
```

8.2.3.21 Intersects (BoundingBox box)

Intersects (BoundingBox box) 表示检查当前点对象是否在指定的边界矩形框内。这里调用 BoundingBox 的 Contains() 方法进行相交检查，代码如下：

```
public bool Intersects(BoundingBox box)
{
    return box.Contains(this);
}
```

8.2.3.22 Contains (Geometry geom)

Contains (Geometry geom) 表示检查当前点对象是否包含指定的几何对象。这里直接返回 False，因为点对象不可能再包含其他的对象，代码如下：

```
public override bool Contains(Geometry geom)
{
    return false;
}
```

8.3 复合点对象 MultiPoint

MultiPoint 是一个复合几何对象，由多个点组成。

8.3.1 MultiPoint 的属性

8.3.1.1 this [int n]

this [int n] 表示通过索引值得到一个点，代码如下：

```
public new Point this[int n]
{
    get { return _Points[n]; }
}
```

8.3.1.2 Points

Points 表示点集对象。这是点的集合，装在一系列点对象的容器中，代码如下：

```
private IList<Point> _Points;
public IList<Point> Points
{
    get { return _Points; }
    set { _Points = value; }
}
```

8.3.1.3 NumGeometries

NumGeometries 表示返回该对象中几何对象的数目。这里就是所装载的点对象的数量，代码如下：

```
public override int NumGeometries
{
    get { return _Points.Count; }
}
```

8.3.1.4 Dimension

Dimension 表示当前几何对象的维度。这里直接返回 0，代码如下：

```
public override int Dimension
{
    get { return 0; }
}
```

8.3.1.5 GeometryType

GeometryType 表示当前几何对象的类型。这里使用的是 GeometryType2. MultiPoint 枚举值，表示复合点对象，代码如下：

```
public override GeometryType2 GeometryType
{
    get
    {
        return GeometryType2.MultiPoint;
    }
}
```

8.3.2 MultiPoint 的方法

8.3.2.1 MultiPoint()

此为构造函数。这里只是实例化了一个点集的容器对象 Collection < Point >。代码如下：

```
public MultiPoint()
{
    _Points = new Collection<Point>();
}
```

8.3.2.2 MultiPoint(IEnumerable < double [ ] > points)

此为构造函数。这里使用 double 数组的迭代对象进行创建。代码如下：

```
public MultiPoint(IEnumerable<double[]> points)
{
    _Points = new Collection<Point>();
    foreach (double[] point in points)
        _Points.Add(new Point(point[0], point[1]));
}
```

8.3.2.3 Geometry(int N)

Geometry(int N) 表示通过索引值得到点对象，代码如下：

```
public new Point Geometry(int N)
{
    return _Points[N];
}
```

8.3.2.4 IsEmpty()

IsEmpty() 表示设置当前复合对象为空，代码如下：

```
public override bool IsEmpty()
```

```
            return (_Points ! = null && _Points.Count = = 0);
    }
```

8.3.2.5 IsSimple()

IsSimple()表示当前的复合对象是否为简单对象,代码如下:

```
    public override bool IsSimple()
    {
            throw new NotImplementedException();
    }
```

8.3.2.6 Boundary()

Boundary()表示复合对象的边界。这里返回空,代码如下:

```
    public override Geometry Boundary()
    {
            return null;
    }
```

8.3.2.7 Distance(Geometry geom)

Distance(Geometry geom)表示当前复合对象与指定的几何对象之间的距离。这里并未实现,代码如下:

```
    public override double Distance(Geometry geom)
    {
            throw new NotImplementedException();
    }
```

8.3.2.8 Buffer(double d)

Buffer(double d)表示复合对象在指定距离的缓冲几何对象,代码如下:

```
    public override Geometry Buffer(double d)
    {
            throw new NotImplementedException();
    }
```

8.3.2.9 ConvexHull()

ConvexHull()表示复合对象的凸壳。这里也未实现,代码如下:

```
    public override Geometry ConvexHull()
    {
            throw new NotImplementedException();
    }
```

8.3.2.10 Intersection(Geometry geom)

Intersection(Geometry geom)表示复合对象与指定的几何对象相交的几何对象。目前

8.3 复合点对象 MultiPoint

版本中尚未实现，代码如下：

```
public override Geometry Intersection(Geometry geom)
{
    throw new NotImplementedException();
}
```

8.3.2.11　Union（Geometry geom）

复合对象与指定的几何对象的并集组成的几何对象。目前版本中尚未实现，代码如下：

```
public override Geometry Union(Geometry geom)
{
    throw new NotImplementedException();
}
```

8.3.2.12　Difference（Geometry geom）

复合对象与指定的几何对象的不同部分组成的几何对象。目前版本中尚未实现，代码如下：

```
public override Geometry Difference(Geometry geom)
{
    throw new NotImplementedException();
}
```

8.3.2.13　SymDifference（Geometry geom）

复合对象与指定的几何对象的对称不同部分组成的几何对象。目前版本中尚未实现，代码如下：

```
public override Geometry SymDifference(Geometry geom)
{
    throw new NotImplementedException();
}
```

8.3.2.14　GetBoundingBox（）

得到当前复合对象的边界矩形框。这里遍历复合对象中的点的集合，将边界矩形设置为点集中的边界值，代码如下：

```
public override BoundingBox GetBoundingBox()
{
    if (_Points == null || _Points.Count == 0)
        return null;
    BoundingBox bbox = new BoundingBox(_Points[0], _Points[0]);
    for (int i = 1; i < _Points.Count; i++)
    {
        bbox.Min.X = _Points[i].X < bbox.Min.X ? _Points[i].X : bbox.Min.X;
```

```
                bbox.Min.Y = _Points[i].Y < bbox.Min.Y ? _Points[i].Y : bbox.Min.Y;
                bbox.Max.X = _Points[i].X > bbox.Max.X ? _Points[i].X : bbox.Max.X;
                bbox.Max.Y = _Points[i].Y > bbox.Max.Y ? _Points[i].Y : bbox.Max.Y;
        }
        return bbox;
}
```

8.3.2.15 Clone()

复合对象的拷贝。这里将点集中的每个点都拷贝后返回，代码如下：

```
public new MultiPoint Clone()
{
        MultiPoint geoms = new MultiPoint();
        for (int i = 0; i < _Points.Count; i++)
                geoms.Points.Add(_Points[i].Clone());
        return geoms;
}
```

8.3.2.16 GetEnumerator()

得到复合对象中的点的迭代，代码如下：

```
public override IEnumerator<Geometry> GetEnumerator()
{
        foreach (Point p in _Points)
                yield return p;
}
```

8.4 线状几何形状的抽象类 Curve

8.4.1 Curve 的属性

8.4.1.1 Dimension

表示曲线的维度。这里直接返回1，代码如下：

```
public override int Dimension
{
        get { return 1; }
}
```

8.4.1.2 Length

返回曲线的长度，代码如下：

```
public abstract double Length { get; }
```

8.4.1.3 StartPoint

返回曲线的起始点，代码如下：

```
public abstract Point StartPoint { get; }
```

8.4.1.4 EndPoint

返回曲线的终止点,代码如下:

```
public abstract Point EndPoint { get; }
```

8.4.1.5 IsClosed

表示曲线是否闭合。当闭合时,终止点就是起始点,它们使用 Equals() 方法会返回 True,代码如下:

```
public bool IsClosed
{
    get { return (StartPoint.Equals(EndPoint)); }
}
```

8.4.1.6 IsRing

线是否为环状,代码如下:

```
public abstract bool IsRing { get; }
```

8.4.1.7 GeometryType

当前曲线的几何类型。这里使用 GeometryType2.Curve 枚举值,表示是曲线类型,代码如下:

```
public override GeometryType2 GeometryType
{
    get
    {
        return GeometryType2.Curve;
    }
}
```

8.4.2 Curve 的方法

Value(double t) 是曲线抽象类的唯一方法,它可以得到指定长度的点的坐标,代码如下:

```
public abstract Point Value(double t);
```

8.5 多边形 Polygon

8.5.1 Polygon 的属性

8.5.1.1 ExteriorRing

外边框线,用一个 LinearRing 类型的线来表示,代码如下:

```csharp
private LinearRing _ExteriorRing;
public LinearRing ExteriorRing
{
    get { return _ExteriorRing; }
    set { _ExteriorRing = value; }
}
```

8.5.1.2 InteriorRings

内部的边线，使用一个 IList < LinearRing > 容器来表示，代码如下：

```csharp
private IList < LinearRing > _InteriorRings;
public IList < LinearRing > InteriorRings
{
    get { return _InteriorRings; }
    set { _InteriorRings = value; }
}
```

8.5.1.3 NumInteriorRing

内部边线的数量，是一个 int 类型的数值，代码如下：

```csharp
public int NumInteriorRing
{
    get { return _InteriorRings.Count; }
}
```

8.5.1.4 Area

多边形的面积。这里面积的计算是外边线的面积减去所有顺时针内边线的面积，加上所有逆时针内边线的面积，代码如下：

```csharp
public override double Area
{
    get
    {
        double area = 0.0;
        area += _ExteriorRing.Area;
        bool extIsClockwise = _ExteriorRing.IsCCW();
        for (int i = 0; i < _InteriorRings.Count; i++)
            if (_InteriorRings[i].IsCCW() != extIsClockwise)
                area -= _InteriorRings[i].Area;
            else
                area += _InteriorRings[i].Area;
        return area;
    }
}
```

8.5 多边形 Polygon

8.5.1.5 Centroid

多边形的数学中心。这里直接使用外边线外包框的中心点作为多边形的数学中心，代码如下：

```
public override Point Centroid
{
    get { return ExteriorRing.GetBoundingBox().GetCentroid(); }
}
```

8.5.1.6 PointOnSurface

保证一个点必须在此多边形内。这里并未实现，代码如下：

```
public override Point PointOnSurface
{
    get { throw new NotImplementedException(); }
}
```

8.5.1.7 GeometryType

当前多边形的几何类型。这里使用的是 GeometryType2.Polygon 枚举值，表示是多边形类型，代码如下：

```
public override GeometryType2 GeometryType
{
    get
    {
        return GeometryType2.Polygon;
    }
}
```

8.5.2 Polygon 的方法

8.5.2.1 Polygon（LinearRing exteriorRing，IList < LinearRing > interiorRings）

此为构造函数。这里使用指定的外边界线和内边界线的集合来初始化一个新的多边形对象，代码如下：

```
public Polygon(LinearRing exteriorRing, IList < LinearRing > interiorRings)
{
    _ExteriorRing = exteriorRing;
    _InteriorRings = interiorRings ?? new Collection < LinearRing > ();
}
```

8.5.2.2 Polygon（LinearRing exteriorRing）

此为构造函数。这里使用指定的外边界线来实例化一个多边形对象，代码如下：

```
public Polygon(LinearRing exteriorRing) : this(exteriorRing, new Collection < LinearRing > ())
{ }
```

8.5.2.3 Polygon()

默认的构造函数,实例化一个空的多边形对象,代码如下:

```
public Polygon() : this(new LinearRing(), new Collection < LinearRing > ())
{}
```

8.5.2.4 InteriorRing(int N)

通过索引值得到相应的内边界线,代码如下:

```
public LinearRing InteriorRing(int N)
{
    return _InteriorRings[N];
}
```

8.5.2.5 TransformToImage(Map map)

将多边形对象转换为指定的地图坐标。在此方法内部,首先生成一个空多边形内的顶点数组,对每个多边形的顶点都使用 Transform.WorldtoMap() 方法进行坐标系统的变换,将转换的结果加入节点集合 v 中;然后获取外环的个数,遍历每个外环的节点,将节点转换为地图坐标,并加入节点集合 v 中,转换完成后,返回此节点集合 v,代码如下:

```
public PointF[] TransformToImage(Map map)
{
    int vertices = _ExteriorRing.Vertices.Count;
    for (int i = 0; i < _InteriorRings.Count; i++)
        vertices += _InteriorRings[i].Vertices.Count;
    PointF[] v = new PointF[vertices];
    for (int i = 0; i < _ExteriorRing.Vertices.Count; i++)
        v[i] = Transform.WorldtoMap(_ExteriorRing.Vertices[i], map);
    int j = _ExteriorRing.Vertices.Count;
    for (int k = 0; k < _InteriorRings.Count; k++)
    {
        for (int i = 0; i < _InteriorRings[k].Vertices.Count; i++)
            v[j + i] = Transform.WorldtoMap(_InteriorRings[k].Vertices[i], map);
        j += _InteriorRings[k].Vertices.Count;
    }
    return v;
}
```

8.5.2.6 GetBoundingBox()

返回多边形的边界矩形。若外环为空,或外环的节点数为 0,返回 null;若不为空,新建一个 BoundingBox 对象 bbox,遍历外环的每个节点,计算出节点 X 和 Y 坐标的最小和最大值,设置 bbox 的 Min 和 Y 属性,代码如下:

```
public override BoundingBox GetBoundingBox()
{
```

8.5 多边形 Polygon

```
    if (_ExteriorRing = = null || _ExteriorRing.Vertices.Count = = 0) return null;
    BoundingBox bbox = new BoundingBox(_ExteriorRing.Vertices[0], _ExteriorRing.Vertices[0]);
    for (int i = 1; i < _ExteriorRing.Vertices.Count; i + +)
    {
        bbox.Min.X = Math.Min(_ExteriorRing.Vertices[i].X, bbox.Min.X);
        bbox.Min.Y = Math.Min(_ExteriorRing.Vertices[i].Y, bbox.Min.Y);
        bbox.Max.X = Math.Max(_ExteriorRing.Vertices[i].X, bbox.Max.X);
        bbox.Max.Y = Math.Max(_ExteriorRing.Vertices[i].Y, bbox.Max.Y);
    }
    return bbox;
}
```

8.5.2.7 Clone()

多边形拷贝。这里先新建一个对变形对象 p，不仅要拷贝其外环，设置外环属性；对于内环，也要对内环集合进行遍历，逐个进行拷贝，设置多边形 p 的内环属性集合，代码如下：

```
public new Polygon Clone()
{
    Polygon p = new Polygon();
    p.ExteriorRing = (LinearRing)_ExteriorRing.Clone();
    for (int i = 0; i < _InteriorRings.Count; i + +)
        p.InteriorRings.Add(_InteriorRings[i].Clone() as LinearRing);
    return p;
}
```

8.5.2.8 Equals(Polygon p)

当前多边形和指定的多边形是否相等，代码如下：

```
public bool Equals(Polygon p)
{
    if (p = = null)
        return false;
    if (!p.ExteriorRing.Equals(ExteriorRing))
        return false;
    if (p.InteriorRings.Count ! = InteriorRings.Count)
        return false;
    for (int i = 0; i < p.InteriorRings.Count; i + +)
        if (!p.InteriorRings[i].Equals(InteriorRings[i]))
            return false;
    return true;
}
```

8.5.2.9 GetHashCode ( )

返回当前多边形的哈希值,代码如下:

```
public override int GetHashCode()
{
    int hash = ExteriorRing.GetHashCode();
    for (int i = 0; i < InteriorRings.Count; i++)
        hash = hash ^ InteriorRings[i].GetHashCode();
    return hash;
}
```

8.5.2.10 IsEmpty ( )

当前多边形是否为空,代码如下:

```
public override bool IsEmpty()
{
    return (ExteriorRing == null) || (ExteriorRing.Vertices.Count == 0);
}
```

8.5.2.11 IsSimple ( )

当前多边形是否为简单多边形。这里并未实现,代码如下:

```
public override bool IsSimple()
{
    throw new NotImplementedException();
}
```

8.5.2.12 Boundary ( )

当前多边形的边界。这里并未实现,代码如下:

```
public override Geometry Boundary()
{
    throw new NotImplementedException();
}
```

8.5.2.13 Distance (Geometry geom)

当前多边形与指定的几何对象的距离。这里并未实现,代码如下:

```
public override double Distance(Geometry geom)
{
    throw new NotImplementedException();
}
```

8.5.2.14 Buffer (double d)

当前多边形在指定距离的缓冲几何对象。目前并未实现,代码如下:

```
public override Geometry Buffer(double d)
```

8.5 多边形 Polygon

```
    throw new NotImplementedException();
}
```

8.5.2.15 ConvexHull()

当前多边形的凸壳。这里也未实现，代码如下：

```
public override Geometry ConvexHull()
{
    throw new NotImplementedException();
}
```

8.5.2.16 Intersection(Geometry geom)

返回当前多边形与指定的几何对象是否相交。这里并未实现，代码如下：

```
public override Geometry Intersection(Geometry geom)
{
    throw new NotImplementedException();
}
```

8.5.2.17 Union(Geometry geom)

返回当前多边形与指定的几何对象的并集。这里并未实现，代码如下：

```
public override Geometry Union(Geometry geom)
{
    throw new NotImplementedException();
}
```

8.5.2.18 Difference(Geometry geom)

返回当前多边形与指定的几何对象不同部分的几何对象。这里并未实现，代码如下：

```
public override Geometry Difference(Geometry geom)
{
    throw new NotImplementedException();
}
```

8.5.2.19 SymDifference(Geometry geom)

返回当前多边形与指定的几何对象对称不同部分的几何对象。这里并未实现，代码如下：

```
public override Geometry SymDifference(Geometry geom)
{
    throw new NotImplementedException();
}
```

8.6 外包矩形框 BoundingBox

8.6.1 BoundingBox 的属性

8.6.1.1 Min
表示边界矩形的左下角点，代码如下：

```
private Point _Min;
public Point Min
{
    get { return _Min; }
    set { _Min = value; }
}
```

8.6.1.2 Max
表示边界矩形的右上角点，代码如下：

```
private Point _Max;
public Point Max
{
    get { return _Max; }
    set { _Max = value; }
}
```

8.6.1.3 Left
返回左边界的值，就是左下角点的 X 坐标值，代码如下：

```
public Double Left
{
    get { return _Min.X; }
}
```

8.6.1.4 Right
返回右边界的值，就是右上角点的 X 坐标值，代码如下：

```
public Double Right
{
    get { return _Max.X; }
}
```

8.6.1.5 Top
返回上边界的值，就是右上角点的 Y 坐标值，代码如下：

```
public Double Top
{
```

8.6 外包矩形框 BoundingBox

```
        get { return _Max.Y; }
}
```

8.6.1.6 Bottom
返回底部边界的值,就是左下角点的 Y 坐标值,代码如下:

```
public Double Bottom
{
        get { return _Min.Y; }
}
```

8.6.1.7 TopLeft
返回左上角的点对象,代码如下:

```
public Point TopLeft
{
        get { return new Point(Left, Top); }
}
```

8.6.1.8 TopRight
返回右上角的点对象,代码如下:

```
public Point TopRight
{
        get { return new Point(Right, Top); }
}
```

8.6.1.9 BottomLeft
返回左下角的点对象,代码如下:

```
public Point BottomLeft
{
        get { return new Point(Left, Bottom); }
}
```

8.6.1.10 BottomRight
返回右下角的点对象,代码如下:

```
public Point BottomRight
{
        get { return new Point(Right, Bottom); }
}
```

8.6.1.11 Width
返回边界矩形的宽度,代码如下:

```
public double Width
{
```

8.6.1.12 Height

返回边界矩形的高度，代码如下：

```
public double Height
{
    get { return Math.Abs(_Max.Y - _Min.Y); }
}
```

8.6.1.13 LongestAxis

返回边界矩形的较长边的索引（在各个索引值代表了点的 X、Y，值为 0 表示 X，值为 1 表示 Y），代码如下：

```
public uint LongestAxis
{
    get
    {
        Point boxdim = Max - Min;
        uint la = 0; // longest axis
        double lav = 0; // longest axis length
        for (uint ii = 0; ii < 2; ii++)
        {
            if (boxdim[ii] > lav)
            {
                la = ii;
                lav = boxdim[ii];
            }
        }
        return la;
    }
}
```

8.6.2 BoundingBox 的方法

8.6.2.1 BoundingBox (double minX, double minY, double maxX, double maxY)

此为构造函数。用指定的坐标值生成一个新的边界矩形，需要调用 CheckMinMax ( ) 方法检查传入的坐标值是否能构成一个矩形，代码如下：

```
public BoundingBox(double minX, double minY, double maxX, double maxY)
{
    _Min = new Point(minX, minY);
    _Max = new Point(maxX, maxY);
    CheckMinMax();
}
```

8.6.2.2 BoundingBox (Point lowerLeft, Point upperRight)

此为构造函数。用指定的坐标点实例化一个边界矩形,代码如下:

```
public BoundingBox(Point lowerLeft, Point upperRight)
    : this(lowerLeft.X, lowerLeft.Y, upperRight.X, upperRight.Y)
{
}
```

8.6.2.3 BoundingBox (Collection < Geometry > objects)

此为构造函数。用指定的几何对象的集合,生成它们的边界矩形。这里遍历每个几何对象,将几何对象的边界值赋予这个边界对象,代码如下:

```
public BoundingBox(Collection < Geometry > objects)
{
    if (objects == null || objects.Count == 0)
    {
        _Min = null;
        _Max = null;
        return;
    }
    _Min = objects[0].GetBoundingBox().Min.Clone();
    _Max = objects[0].GetBoundingBox().Max.Clone();
    CheckMinMax();
    for (int i = 1; i < objects.Count; i++)
    {
        BoundingBox box = objects[i].GetBoundingBox();
        _Min.X = Math.Min(box.Min.X, Min.X);
        _Min.Y = Math.Min(box.Min.Y, Min.Y);
        _Max.X = Math.Max(box.Max.X, Max.X);
        _Max.Y = Math.Max(box.Max.Y, Max.Y);
    }
}
```

8.6.2.4 BoundingBox (Collection < BoundingBox > objects)

此为构造函数。这个构造函数使用边界矩形的集合来构造出它们的边界矩形,构造方法同上,代码如下:

```
public BoundingBox(Collection < BoundingBox > objects)
{
    if (objects.Count == 0)
    {
        _Max = null;
        _Min = null;
```

```
        }
        else
        {
            _Min = objects[0].Min.Clone();
            _Max = objects[0].Max.Clone();
            for (int i = 1; i < objects.Count; i++)
            {
                _Min.X = Math.Min(objects[i].Min.X, Min.X);
                _Min.Y = Math.Min(objects[i].Min.Y, Min.Y);
                _Max.X = Math.Max(objects[i].Max.X, Max.X);
                _Max.Y = Math.Max(objects[i].Max.Y, Max.Y);
            }
        }
    }
```

8.6.2.5 Offset(Point vector)

边界矩形的偏移值。这里仅需将边界矩形的两个边界点分别"加上"偏移值就可以达到偏移的效果,代码如下:

```
    public void Offset(Point vector)
    {
        _Min += vector;
        _Max += vector;
    }
```

8.6.2.6 CheckMinMax()

检查左下角和右上角能否构成一个正矩形,代码如下:

```
    public bool CheckMinMax()
    {
        bool wasSwapped = false;
        if (_Min.X > _Max.X)
        {
            double tmp = _Min.X;
            _Min.X = _Max.X;
            _Max.X = tmp;
            wasSwapped = true;
        }
        if (_Min.Y > _Max.Y)
        {
            double tmp = _Min.Y;
            _Min.Y = _Max.Y;
            _Max.Y = tmp;
            wasSwapped = true;
```

```
        return wasSwapped;
    }
```

8.6.2.7　Intersects（BoundingBox box）

判断当前边界矩形是否和指定的边界矩形相交。这里只需要判断它们的边界是否相交，速度很快，很多时候两个几何对象相交的判断也是将其转换为各自的边界矩形的相交判断，代码如下：

```
public bool Intersects(BoundingBox box)
{
    return !(box.Min.X > Max.X ||
             box.Max.X < Min.X ||
             box.Min.Y > Max.Y ||
             box.Max.Y < Min.Y);
}
```

8.6.2.8　Intersects（Geometry g）

判断当前边界矩形是否和指定的几何对象相交。这里直接调用Touches（）方法进行相交判断，代码如下：

```
public bool Intersects(Geometry g)
{
    return Touches(g);
}
```

8.6.2.9　Touches（BoundingBox r）

判断当前边界矩形是否和指定的边界矩形接触。这里判断它们的边界值是否相交，速度也很快，代码如下：

```
public bool Touches(BoundingBox r)
{
    for (uint cIndex = 0; cIndex < 2; cIndex++)
    {
        if ((Min[cIndex] > r.Min[cIndex] && Min[cIndex] < r.Min[cIndex]) ||
            (Max[cIndex] > r.Max[cIndex] && Max[cIndex] < r.Max[cIndex]))
            return true;
    }
    return false;
}
```

8.6.2.10　Touches（Geometry s）

Touches（）方法的重载版本。这里仅仅支持点对象的判断，其他类型对象的判断这里会抛出未实现异常，代码如下：

```
public bool Touches(Geometry s)
```

```
        if (s is Point) return Touches(s as Point);
        throw new NotImplementedException("Touches: Not implemented on this geometry type");
    }
```

8.6.2.11　Contains（BoundingBox r）

当前边界矩形是否包含指定的边界矩形，若包含则返回 True，代码如下：

```
    public bool Contains(BoundingBox r)
    {
        for (uint cIndex = 0; cIndex < 2; cIndex++)
            if (Min[cIndex] > r.Min[cIndex] || Max[cIndex] < r.Max[cIndex]) return false;
        return true;
    }
```

8.6.2.12　Contains（Geometry s）

Contains 方法的重载版本，这里也只能对点对象矩形判断，代码如下：

```
    public bool Contains(Geometry s)
    {
        if (s is Point) return Contains(s as Point);
        throw new NotImplementedException("Contains: Not implemented on these geometries");
    }
```

8.6.2.13　Touches（Point p）

判断当前的边界矩形是否和指定的点对象接触，代码如下：

```
    public bool Touches(Point p)
    {
        for (uint cIndex = 0; cIndex < 2; cIndex++)
        {
            if ((Min[cIndex] > p[cIndex] && Min[cIndex] < p[cIndex]) ||
                (Max[cIndex] > p[cIndex] && Max[cIndex] < p[cIndex]))
                return true;
        }
        return false;
    }
```

8.6.2.14　GetArea（）

返回边界矩形的面积即长度×宽度，代码如下：

```
    public double GetArea()
    {
        return Width * Height;
    }
```

8.6.2.15 GetIntersectingArea（BoundingBox r）

返回和指定边界矩形相交部分的面积，代码如下：

```
public double GetIntersectingArea(BoundingBox r)
{
    uint cIndex;
    for (cIndex = 0; cIndex < 2; cIndex + +)
        if (Min[cIndex] > r.Max[cIndex] || Max[cIndex] < r.Min[cIndex]) return 0.0;
    double ret = 1.0;
    double f1, f2;
    for (cIndex = 0; cIndex < 2; cIndex + +)
    {
        f1 = Math.Max(Min[cIndex], r.Min[cIndex]);
        f2 = Math.Min(Max[cIndex], r.Max[cIndex]);
        ret *= f2 - f1;
    }
    return ret;
}
```

8.6.2.16 Join（BoundingBox box）

返回当前边界矩形和指定的边界矩形并集的边界矩形，代码如下：

```
public BoundingBox Join(BoundingBox box)
{
    if (box == null)
        return Clone();
    else
        return new BoundingBox(Math.Min(Min.X, box.Min.X), Math.Min(Min.Y, box.Min.Y), Math.Max(Max.X, box.Max.X), Math.Max(Max.Y, box.Max.Y));
}
```

8.6.2.17 Join（BoundingBox box1，BoundingBox box2）

返回指定的第一个边界矩形和第二个边界矩形的并集的边界矩形，代码如下：

```
public static BoundingBox Join(BoundingBox box1, BoundingBox box2)
{
    if (box1 == null && box2 == null)
        return null;
    else if (box1 == null)
        return box2.Clone();
    else
        return box1.Join(box2);
}
```

8.6.2.18 Join（BoundingBox [ ] boxes）

返回指定的一系列边界矩形的并集的边界矩形，代码如下：

```csharp
public static BoundingBox Join(BoundingBox[ ] boxes)
{
    if (boxes = = null) return null;
    if (boxes.Length = = 1) return boxes[0];
    BoundingBox box = boxes[0].Clone();
    for (int i = 1; i < boxes.Length; i + +)
        box = box.Join(boxes[i]);
    return box;
}
```

8.6.2.19　Grow（double amount）

使当前的边界矩形"增大"指定的值,这个值为正数时会有放大的效果,为负值时会有缩小的效果,当然,执行操作后需要检查边界点还能否构成合法的边界矩形,代码如下:

```csharp
public BoundingBox Grow(double amount)
{
    BoundingBox box = Clone();
    box.Min.X - = amount;
    box.Min.Y - = amount;
    box.Max.X + = amount;
    box.Max.Y + = amount;
    box.CheckMinMax();
    return box;
}
```

8.6.2.20　Grow（double amountInX, double amountInY）

Grow 的重载方法。这里分别给 X、Y"增加"不同的值,代码如下:

```csharp
public BoundingBox Grow(double amountInX, double amountInY)
{
    BoundingBox box = Clone();
    box.Min.X - = amountInX;
    box.Min.Y - = amountInY;
    box.Max.X + = amountInX;
    box.Max.Y + = amountInY;
    box.CheckMinMax();
    return box;
}
```

8.6.2.21　Contains（Point p）

判断指定的点对象是否在当前的边界矩形中,代码如下:

```csharp
public bool Contains(Point p)
{
```

```
        if ( Max. X  <  p. X)
            return false;
        if ( Min. X  >  p. X)
            return false;
        if ( Max. Y  <  p. Y)
            return false;
        if ( Min. Y  >  p. Y)
            return false;
        return true;
    }
```

8.6.2.22　Distance（BoundingBox box）

计算当前边界矩形和指定的边界矩形的距离。两个相交的边界矩形的距离为 0，其他情况下边界矩形的距离就是最临近点之间的距离，代码如下：

```
    public virtual double Distance( BoundingBox box)
    {
        double ret = 0.0;
        for ( uint cIndex = 0; cIndex < 2; cIndex + + )
        {
            double x = 0.0;
            if ( box. Max[ cIndex] < Min[ cIndex] ) x = Math. Abs( box. Max[ cIndex] − Min[ cIndex] );
            else if ( Max[ cIndex] < box. Min[ cIndex] ) x = Math. Abs( box. Min[ cIndex] − Max[ cIndex] );
            ret + = x * x;
        }
        return Math. Sqrt( ret);
    }
```

8.6.2.23　Distance（Point p）

当前边界矩形和指定的点之间的距离，代码如下：

```
    public virtual double Distance( Point p)
    {
        double ret = 0.0;
        for ( uint cIndex = 0; cIndex < 2; cIndex + + )
        {
            if ( p[ cIndex] < Min[ cIndex] ) ret + = Math. Pow( Min[ cIndex] − p[ cIndex], 2.0);
            else if ( p[ cIndex] > Max[ cIndex] ) ret + = Math. Pow( p[ cIndex] − Max[ cIndex], 2.0);
        }
        return Math. Sqrt( ret);
    }
```

8.6.2.24 GetCentroid ( )

返回当前边界矩形的中心,代码如下:

```
public Point GetCentroid( )
{
    return (_Min + _Max) * .5f;
}
```

8.6.2.25 Clone ( )

当前边界矩形的拷贝。这里直接生成一个全新的边界矩形对象 BoundingBox,代码如下:

```
public BoundingBox Clone( )
{
    return new BoundingBox(_Min.X, _Min.Y, _Max.X, _Max.Y);
}
```

8.6.2.26 ToString ( )

返回由地图数字格式、地图 X 最小值、地图 Y 最小值、地图 X 最大值、地图 Y 最大值组成的字符串,代码如下:

```
public override string ToString( )
{
    return String.Format(Map.NumberFormatEnUs, "{0}, {1} {2}, {3}", Min.X, Min.Y, Max.X, Max.Y);
}
```

8.6.2.27 Equals (object obj)

当前边界矩形与制定对象的相等性的比较。对于不是边界矩形的对象直接返回 False,是边界矩形的对象则调用 Equals ( ) 重载方法进行判断,代码如下:

```
public override bool Equals(object obj)
{
    BoundingBox box = obj as BoundingBox;
    if (obj == null) return false;
    else return Equals(box);
}
```

8.6.2.28 GetHashCode ( )

得到当前边界矩形的哈希值,代码如下:

```
public override int GetHashCode( )
{
    return Min.GetHashCode( ) ^ Max.GetHashCode( );
}
```

8.7 空间关系类 SpatialRelations

用于封装判断空间关系的类，其中 Intersects、Overlaps、Touches 方法未实现。

（1）Contains（Geometry sourceGeometry，Geometry otherGeometry）。此方法用于判断指定的几何对象是否包含另一个几何对象。这里直接调用源几何对象的 Within（）方法进行包含关系判断，代码如下：

```
public static bool Contains(Geometry sourceGeometry, Geometry otherGeometry)
{
    return (otherGeometry.Within(sourceGeometry));
}
```

（2）Crosses（Geometry g1，Geometry g2）。判断两个几何对象是否交叉。这里先对其实行相交判断，得到相交的几何对象，若这个相交集合对象的维度小于这两个几何对象的最大值，且这两个几何对象不相等，则返回 true，代码如下：

```
public static bool Crosses(Geometry g1, Geometry g2)
{
    Geometry g = g2.Intersection(g1);
    return (g.Intersection(g1).Dimension < Math.Max(g1.Dimension, g2.Dimension) && !g.Equals(g1) &&
        !g.Equals(g2));
}
```

（3）Disjoint（Geometry g1，Geometry g2）。返回两个几何对象的相交几何对象，这里直接调用几何对象的 Intersects（）方法得到相交判断后的几何对象并返回，代码如下：

```
public static bool Disjoint(Geometry g1, Geometry g2)
{
    return !g2.Intersects(g1);
}
```

（4）Equals（Geometry g1，Geometry g2）。将两个几何对象进行相等性比较，这里需要针对不同的几何类型的对象分别处理，调用各自的 Equals（）方法进行判断，代码如下：

```
public static bool Equals(Geometry g1, Geometry g2)
{
    if (g1 == null && g2 == null)
        return true;
    if (g1 == null || g2 == null)
        return false;
```

```
            if ( g1. GetType( ) ！ = g2. GetType( ) )
                return false；
            if ( g1 is Point)
                return ( g1 as Point). Equals( g2 as Point)；
            if ( g1 is LineString)
                return ( g1 as LineString). Equals( g2 as LineString)；
            if ( g1 is Polygon)
                return ( g1 as Polygon). Equals( g2 as Polygon)；
            if ( g1 is MultiPoint)
                return ( g1 as MultiPoint). Equals( g2 as MultiPoint)；
            if ( g1 is MultiLineString)
                return ( g1 as MultiLineString). Equals( g2 as MultiLineString)；
            if ( g1 is MultiPolygon)
                return ( g1 as MultiPolygon). Equals( g2 as MultiPolygon)；
            throw new ArgumentException("The method or operation is not implemented on this geometry
                type. ")；
        }
```

（5）Intersects（Geometry g1，Geometry g2）。判断两个几何对象是否相交。这里并未实现，代码如下：

```
        public static bool Intersects( Geometry g1，Geometry g2)
        {
            throw new NotImplementedException( )；
        }
```

（6）Overlaps（Geometry g1，Geometry g2）。判断两个几何对象是否叠加。这里并未实现，代码如下：

```
        public static bool Overlaps( Geometry g1，Geometry g2)
        {
            throw new NotImplementedException( )；
        }
```

（7）Touches（Geometry g1，Geometry g2）。判断两个几何对象是否接触。这里并未实现，代码如下：

```
        public static bool Touches( Geometry g1，Geometry g2)
        {
            throw new NotImplementedException( )；
        }
```

（8）Within（Geometry g1，Geometry g2）。判断指定的源几何对象（第一个参数）是否包含目标几何对象（第二个参数）。这里调用几何对象的 Contains（ ）方法来进行判断，代码如下：

```
public static bool Within( Geometry g1, Geometry g2)
{
    return g1.Contains(g2);
}
```

复习思考题

8-1 SharpMap 支持的几何对象有哪些？若有 SharpMap 不支持的几何对象，思考如何对其进行扩展。

8-2 试编程实现对两个几何对象相交、包含等空间关系的判断。

第 9 章 Windows 应用程序开发
——WinFormSamples

本书第一部分（1~8 章）详细解释了 SharpMap 类库中的主要对象，这一部分（9、10 章）将根据 SharpMap 自带的两个实例，说明 SharpMap 类库的用法。SharpMap 下载包中，有三个例子，其中两个是 Windows 应用程序，一个是 Web 应用程序。Windows 应用程序的项目名称是 DemoWinForm 和 WinFormSamples；Web 应用程序的项目名称是 DemoWebSite。由于篇幅关系，本书未对 Web 应用程序进行详解，请读者参看源代码。

本示例程序用来演示如何使用不同数据源，项目名称为 WinFormSamples。

9.1 数　据

该示例程序所使用的数据在该项目目录下的 GeoData 目录下，共有 8 个目录，76 个文件。使用数据类型如下：

（1）Shape 格式。该格式由 ESRI 提出，可能是当前最流行的 GIS 数据格式。

（2）WMS（Web Map Service）服务（网络地图服务）。WMS 是常见的一种 OGC 地图服务，它将空间数据库中的数据生成图片，发给客户使用。本例中使用的数据来自 http://www2.demis.nl/worldmap/wms.asp，这是一个免费的世界地图数据。

（3）WFS（Web Feature Service）服务（网络要素服务）。WMS 将空间数据的图片发给客户使用，而 WFS 则可将空间数据直接发给客户机器使用。由于通常没有免费的 WFS，本例中系统使用的是 GeoServer（见附录中条目 B），必须安装配置 GeoServer 及相应数据服务，才能访问该数据。

（4）GDAL 数据。GDAL（Geospatial Data Abstraction Library）是一个在 X/MIT 许可协议下的开源栅格空间数据转换库。它利用抽象数据模型来表达所支持的各种文件格式。它还有一系列命令行工具来进行数据转换和处理。本例中，共使用了 8 套栅格数据。

（5）ORG 数据。OGR 是 GDAL 项目的一个分支，它提供对多种矢量数据的支持，如 ESRI、Shapefiles、S-57、SDTS、PostGIS、Oracle Spatial 以及 Mapinfo 等格式。本例中使用了三种数据，分别是 Shapefiles、S-57（见附录条目 C）、Mapinfo。

（6）Tile（瓦片）数据服务。本例中，使用了包含 BingMap 及 GoogleMap 提供的瓦片式数据服务等 9 种地图数据。

（7）PostGIS。PostGIS 是一套开源的空间数据库系统，它底层使用的是 PostgreSQL（见附录 D 条目）数据库系统。在本例中要想使用该数据，必须先安装、配置 PostGIS 数据源。

（8）SpatiaLite。SpatiaLite 是一套小型空间数据库系统（见附录条目 E）。在本例中要想使用该数据，必须先安装、配置 SpatiaLite 数据源。

（9）ORACLE 空间数据。在本例中要想使用该数据，必须先安装、配置 ORACLE 数据源。

9.2 系统简介

系统运行界面如图 9-1 所示。

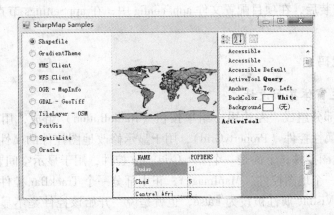

图 9-1 SharpMap Samples 运行界面一

在图 9-1 所示的系统运行界面中，左边的 10 个单选按钮代表 9 种不同的数据源（前两项使用的都是 Shape 格式数据，但 GradienTheme 采用的是渐变色效果的分级符号显示），每种数据源都使用了多套数据，可以通过多次单击达到切换数据的功能，如多次单击第一个按钮 Shapefile，可以发现有两种数据。

在右边的属性窗口中，可通过修改 ActiveTool 属性来改变当前地图的操作工具，图 9-1 是地图工具为 Query（查询）状态下，单击地图要素后，在下面网格控件显示相应要素属性信息的截图。

多次单击 TileLayer 单选按钮，每单击一次会使用不同的数据，当使用数据为 TileLayer-GoogleLabels 时，地图下方会显示出一个 TrackBar 控件，通过该控件可以调整地图的旋转角度。如图 9-2 所示。

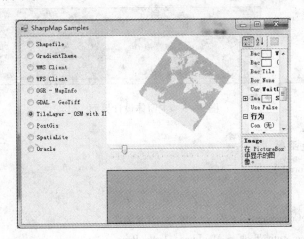

图 9-2 SharpMap Samples 运行界面二

9.3 代码分析

为了使用 GDAL 数据及 OGR 数据，需要在 http://home.gdal.org/fwtools/下载 FW-Tools247.exe，安装后，在项目配置文件 app.config 中，在 appSettings 节点中修改 FWTools 相关路径。

9.3.1 主窗体代码

9.3.1.1 主窗体界面设计

主窗体由 5 部分组成，左边由一组单按钮（RadioButton）组成，用来选取使用的数据；右边是一个属性控件（PropertyGrid），用于显示修改地图控件的属性，其 SelectedObjec 属性被设为地图控件；下面是一个 DataGridView 控件，用于显示空间要素的属性数据；中间部分上面是一个地图控件（MapImage）；正中部分一个 TrackBar 控件，用于控制地图的旋转角度，其 Visible 属性被设为 False，也就是说一开始该控件是不显示的，程序运行当使用 TileLayer 数据时，如果当前显示数据是 TileLayer - GoogleLabels，则设置 TrackBar 控件的 Visible 属性为 True（参见 9.3.1.3 小节内容）。如图 9-3 所示。

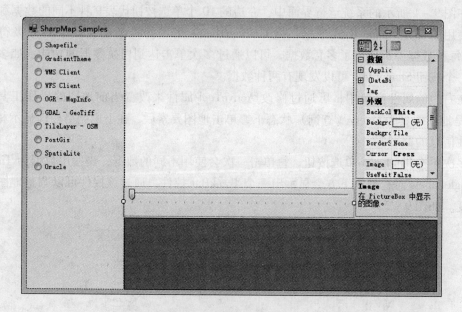

图 9-3 主窗体设计界面

9.3.1.2 构造函数

```
public MainForm()
{
    InitializeComponent();
    mapImage.ActiveTool = MapImage.Tools.Pan;
}
```

构造函数中，将当前地图控件的工具设置为移动。

9.3.1.3 radioButton_Click 函数

radioButton_Click 函数是所有单选按钮的 Click 事件映射函数。

先保存当前地图的鼠标形状，然后将鼠标设置为等待，代码如下：

```
Cursor mic = mapImage.Cursor;
mapImage.Cursor = Cursors.WaitCursor;
Cursor = Cursors.WaitCursor;
```

之后，根据当前单击的单选按钮，执行相应具体访问数据的函数，代码如下：

```
string text = ((RadioButton)sender).Text;
switch (text)
{
    case "Shapefile":
        mapImage.Map = ShapefileSample.InitializeMap();
        break;
    case "GradientTheme":
        mapImage.Map = GradiantThemeSample.InitializeMap();
        break;
    case "WFS Client":
        mapImage.Map = WfsSample.InitializeMap();
        break;
    case "WMS Client":
        mapImage.Map = WmsSample.InitializeMap();
        break;
    case "OGR - MapInfo":
    case "OGR - S-57":
        mapImage.Map = OgrSample.InitializeMap();
        break;
    case "GDAL - GeoTiff":
    case "GDAL - '.DEM'":
    case "GDAL - '.ASC'":
    case "GDAL - '.VRT'":
        mapImage.Map = GdalSample.InitializeMap();
        mapImage.ActiveTool = MapImage.Tools.Pan;
        break;
    case "TileLayer - OSM":
    case "TileLayer - Bing Roads":
    case "TileLayer - Bing Aerial":
    case "TileLayer - Bing Hybrid":
    case "TileLayer - GoogleMap":
    case "TileLayer - GoogleSatellite":
    case "TileLayer - GoogleTerrain":
```

```
            case "TileLayer - GoogleLabels":
                tbAngle.Visible = text.Equals("TileLayer - GoogleLabels");
                if (! tbAngle.Visible) tbAngle.Value = 0;
                mapImage.Map = TileLayerSample.InitializeMap(tbAngle.Value);
                ((RadioButton)sender).Text = mapImage.Map.Layers[0].LayerName;
                break;
            case "PostGis":
                mapImage.Map = PostGisSample.InitializeMap();
                break;
            case "SpatiaLite":
                mapImage.Map = SpatiaLiteSample.InitializeMap();
                break;
            case "Oracle":
                mapImage.Map = OracleSample.InitializeMap();
                break;
            default:
                break;
        }
```

通过以上代码可以看出，当使用 GDAL 数据时，会将地图工具设置为移动（Pan）；当使用 TileLayer 数据时，如果是 TileLayer - GoogleLabels，则显示 TrackBar 控件，从而可以旋转地图。加载新地图数据后，根据新地图数据第一个图层的名称，重新设置单选按钮的 Text 属性。

最后，由于新加载了地图数据，应更新地图尺寸，刷新地图显示，恢复地图鼠标，代码如下：

```
mapImage.Map.Size = Size;
mapImage.Refresh();
Cursor = Cursors.Default;
mapImage.Cursor = mic;
```

9.3.1.4 mapImage_MapQueried 函数

mapImage_MapQueried 函数是地图控件的查寻（Query）事件函数，其将事件函数的传入参数、数据表对象设置为网络控件的数据源，从而显示数据，代码如下：

```
private void mapImage_MapQueried(SharpMap.Data.FeatureDataTable data)
{
    dataGridView1.DataSource = data as System.Data.DataTable;
}
```

9.3.1.5 tbAngle_Scroll 函数及 TrackBar 控件的 Scroll 事件

根据 TrackBar 的值，生成相应的旋转矩阵，再赋值给地图的 MapTransform 属性，代码如下：

```
System.Drawing.Drawing2D.Matrix matrix = new Matrix();
```

```
matrix.RotateAt( tbAngle.Value, new PointF( mapImage.Width * 0.5f, mapImage.Height * 0.5f));
mapImage.Map.MapTransform = matrix;
mapImage.Refresh();
```

9.3.1.6 更新地图属性控件

当地图控件的 ActiveToolChanged、MapCenterChanged、MapRefreshed、MapZoomChanged、MapZooming、SizeChanged 事件发生时，调用 UpdatePropertyGrid 函数以更新地图属性控件（PropertyGrid），代码如下：

```
private void UpdatePropertyGrid()
{
    pgMap.Update();
}
```

9.3.2 数据访问代码

在单选按钮的事件映射函数中，地图是通过调用 ShapefileSample、GradiantThemeSample、WfsSample、WfsSample、OgrSample、GdalSample、PostGisSample、SpatiaLiteSample、OracleSample 类的 InitializeMap 静态函数构建的。以上类存放在项目的 Samples 子目录下。

9.3.2.1 ShapefileSample 类

该数据类支持的是 shape 格式数据，提供两套数据，由一个整型变量控制输出哪一套数据。

A　InitializeMap 函数

当 _mapId 值为 0 时，调用 InitializeMapOrig 函数构造地图对象，并对 _mapId 进行自加操作；否则调用 InitializeMapOsm 函数，并对 _mapId 进行自减操作，代码如下：

```
private static int _mapId = 0;
public static Map InitializeMap()
{
    switch (_mapId)
    {
        case 0:
            _mapId++;
            return InitializeMapOrig();
        case 1:
            _mapId--;
            return InitializeMapOsm();
        default:
            _mapId = 0;
            return InitializeMapOrig();
    }
}
```

B InitializeMapOrig 函数

构造一幅世界地图,共五个图层,包括三个数据层及两个注记层,主要代码解释如下。

首先声名一地图对象;再声名一个矢量图层对象,将其数据源设置为 ShapeFile 对象,数据路径为 GeoData/World/countries.shp;再将图层符号设置为绿色黑边的实心填充符号;最后将投影的 SRID 设置为 4326,关于 SRID 简介请参考附录中的条目 A。

```
//Initialize a new map of size 'imagesize'
Map map = new Map();
//Set up the countries layer
VectorLayer layCountries = new VectorLayer("Countries");
//Set the datasource to a shapefile in the App_data folder
layCountries.DataSource = new ShapeFile("GeoData/World/countries.shp", true);
//Set fill-style to green
layCountries.Style.Fill = new SolidBrush(Color.Green);
//Set the polygons to have a black outline
layCountries.Style.Outline = Pens.Black;
layCountries.Style.EnableOutline = true;
layCountries.SRID = 4326;
```

然后为该图层增加一个注记图层,代码如下:

```
//Set up a country label layer
LabelLayer layLabel = new LabelLayer("Country labels");
layLabel.DataSource = layCountries.DataSource;
layLabel.Enabled = true;
layLabel.LabelColumn = "Name";
layLabel.Style = new LabelStyle();
layLabel.Style.ForeColor = Color.White;
layLabel.Style.Font = new Font(FontFamily.GenericSerif, 12);
layLabel.Style.BackColor = new SolidBrush(Color.FromArgb(128, 255, 0, 0));
layLabel.MaxVisible = 90;
layLabel.MinVisible = 30;
layLabel.Style.HorizontalAlignment = LabelStyle.HorizontalAlignmentEnum.Center;
layLabel.SRID = 4326;
layLabel.MultipartGeometryBehaviour = LabelLayer.MultipartGeometryBehaviourEnum.Largest;
```

上面代码中,声名一 LabelLayer 对象,并将 DataSource(数据源)设置为 layCountries 的数据源。

layLabel.LabelColumn = "Name";表示用 Shape 文件中的 Name 字段为注记内容。之后,设置相应的注记字体、颜色、显示方式、投影编号等。

函数最后将生成的图层加入到地图对象中,并设计显示范围、地图显示中心,代码如下:

```
//Add the layers to the map object.
//The order we add them in are the order they are drawn, so we add the rivers last to put them on top
map.Layers.Add(layCountries);
map.Layers.Add(layRivers);
map.Layers.Add(layCities);
map.Layers.Add(layLabel);
map.Layers.Add(layCityLabel);
//limit the zoom to 360 degrees width
map.MaximumZoom = 360;
map.BackColor = Color.LightBlue;
map.Zoom = 360;
map.Center = new Point(0, 0);
return map;
```

C InitializeMapOsm 函数

相对复杂，通过 partial class（部分类），它被定义于另一个文件（ShapefileSampleOsm.cs）中。该类演示了使用如何通过主题（参见5.6节）实现要素分类显示的功能，程序会根据要素的不同类型，以不同的方式绘制要素，如 natural.shp 数据中，有 forest、water、riverbank、park 几种类型（该信息数据存在 type 字段中），则所有 forest 类型的要素以绿色绘制，water 类型的要素以蓝色绘制。其主要内容解释如下。

首先声名一个 delegate，输入参数为要素对象 FeatureDataRow，返回一个 IStyle 对象：

```
private delegate IStyle GetStyleHandler(FeatureDataRow row);
```

为了达到根据不同类型绘制要素的目的，先构造一个辅助 ITheme（主题）类：

```
class ThemeViaDelegate : ITheme
```

该类有四个数据成员，GetStyleFunction 保存具体 delegate（代理）对象；_default 是默认 IStyle 对象；_columnName 是存放要素类型信息的字段名称；_stylePreserver 是一个 Dictionary（字典容器）对象，用来保存 delegate（代理）对象：

```
public GetStyleHandler GetStyleFunction;
private readonly IStyle _default;
private readonly String _columnName;
private readonly IDictionary<String, IStyle> _stylePreserver;
```

构造函数有两个参数，一个是默认的 IStyle 对象，一个是要素集中存放要素类型信息的字段名称，构造函数将输入参数保存到相关的成员数据中，并生成 Dictionary 对象：

```
public ThemeViaDelegate(IStyle defaultStyle, String columnName)
{
    _default = defaultStyle;
    _stylePreserver = new Dictionary<string, IStyle>();
    _columnName = columnName;
}
```

GetStyle 函数是 ITheme 接口中定义的，实现代码中，首先从要素中读取类型字段的内容（如水系、森林等），再根据类型信息，尝试从_stylePreserver 中获取代理对象，如能得到代理对象，则调用代理对象，得到用于绘制当前要素的 IStyle 对象并返回；否则返回默认 IStyle 对象：

```csharp
public IStyle GetStyle(FeatureDataRow attribute)
{
    IStyle returnStyle;
    String value = Convert.ToString(attribute[_columnName]);
    if (!_stylePreserver.TryGetValue(value, out returnStyle))
    {
        if (GetStyleFunction != null)
        {
            returnStyle = GetStyleFunction(attribute);
            if (returnStyle == null) returnStyle = _default;
            _stylePreserver.Add(value, returnStyle);
        }
        else
            returnStyle = _default;
    }
    return returnStyle;
}
```

有了辅助类后，再看一下构造地图对象的 InitializeMapOsm 函数。

首先构造两个 Style 对象及 Map 对象：

```csharp
VectorStyle transparentStyle = new VectorStyle();
transparentStyle.Fill = Brushes.Transparent;
transparentStyle.EnableOutline = true;//otherwise all the fancy theming stuff won't work!
transparentStyle.Line.Brush = Brushes.Transparent;
transparentStyle.Outline.Brush = Brushes.Transparent;
transparentStyle.Symbol = null;
VectorStyle transparentStyle2 = new VectorStyle();
transparentStyle2.Fill = Brushes.Transparent;
transparentStyle2.EnableOutline = true;//otherwise all the fancy theming stuff won't work!
transparentStyle2.Line.Brush = Brushes.Transparent;
transparentStyle2.Outline.Brush = Brushes.Transparent;
transparentStyle2.Symbol = null;
//Initialize a new map
Map map = new Map();
map.BackColor = Color.Cornsilk;
```

然后构造 5 个图层，以下为配置 natural.shp 代码，其他图层类似，书中不再进行解释。首先构造 VectorLayer 对象，设置数据源及 Style 对象：

```
//Set up the countries layer
VectorLayer layNatural = new VectorLayer("Natural");
//Set the datasource to a shapefile in the App_data folder
layNatural.DataSource = new ShapeFile(string.Format("{0}/natural.shp", PathOsm), true);
//Set default style to draw nothing
layNatural.Style = transparentStyle;
```

关键代码如下：先构造一个 ThemeViaDelegate 对象，然后生成一个匿名函数，并赋值给 ThemeViaDelegate 对象的 GetStyleFunction 属性。代码中，匿名函数先根据读出要素的 type 属性值，再根据其内容配置相应的 VectorStyle 对象，如当 type 等于 forest 时，VectorStyle 对象的 Fill 属性被设为 Brushes.ForestGreen，VectorStyle 对象的 EnableOutline 属性被设为 true，VectorStyle 对象的 Outline.Brush 属性被设为 Brushes.DarkGreen；最后设置图层的 Theme 属性为刚配置好的对象，设置空间参考信息，参见附录条目 A。

```
ThemeViaDelegate theme = new ThemeViaDelegate(layNatural.Style, "type");
theme.GetStyleFunction = delegate(FeatureDataRow row)
{
    string caseVal = (String)row["type"];
    caseVal = caseVal.ToLowerInvariant();
    VectorStyle returnStyle = new VectorStyle();

    switch (caseVal)
    {
        case "forest":
            returnStyle.Fill = Brushes.ForestGreen;
            returnStyle.EnableOutline = true;
            returnStyle.Outline.Brush = Brushes.DarkGreen;
            break;
        case "water":
            returnStyle.Fill = Brushes.Aqua;
            returnStyle.EnableOutline = true;
            returnStyle.Outline.Brush = Brushes.DarkBlue;
            break;
        case "riverbank":
            returnStyle.Fill = Brushes.Peru;
            returnStyle.EnableOutline = true;
            returnStyle.Outline.Brush = Brushes.OrangeRed;
            break;
        case "park":
            returnStyle.Fill = Brushes.PaleGreen;
            returnStyle.EnableOutline = true;
            returnStyle.Outline.Brush = Brushes.DarkGreen;
            break;
```

```
            default:
                returnStyle = null;
                break;
    }
    return returnStyle;
};
layNatural. Theme = theme;
layNatural. SRID = 31466;
```

其余四个图层代码类似,不再列出,值得指出的是,在加载点数据图层中,使用的是项目子目录 Images 中的图片绘制要素。最后,将配置好的图层加到地图对象中去:

```
map. Layers. Add(layNatural);
map. Layers. Add(layWaterways);
map. Layers. Add(layRail);
map. Layers. Add(layRoads);
map. Layers. Add(layPoints);
```

并设置地图对象的最大显示范围、显示范围及地图显示中心:

```
map. MaximumZoom = layRoads. Envelope. Width * 0.75d;
map. Zoom = layRoads. Envelope. Width * 0.2d;
map. Center = layRoads. Envelope. GetCentroid();
```

最后,设置地图的说明文字,在地图右上角显示一行文字:

```
map. Disclaimer = "Geodata from OpenStreetMap (CC - by - SA)\nTransformed to Shapefile by geofabrik. de";
map. DisclaimerFont = new Font("Arial", 7f, FontStyle. Italic);
map. DisclaimerLocation = 1;
```

9.3.2.2 GradiantThemeSample 类

该数据类支持的也是 ShapeFile 格式的数据,提供了一套世界地图数据,包括国家(countries)、城市(cities)、水系(rivers)三个图层,在程序中只加载了前两个图层。数据的加载和注记层的加载同 ShapeFile 数据类相似,这里不再赘述。

数据加载完成后,为了能显示渐变色彩(依据人口密度渐变),需构造相应的 GradientTheme 对象。下面代码构造的 GradientTheme 对象,会有以下行为:当要素 PopDens 属性的值为 0 时,采用 VectorStyle 对象 min 绘制要素;当要素 PopDens 属性的值为 400 时,采用 VectorStyle 对象 max 绘制要素;当要素 PopDens 属性的值位于 0~400 之间时,则通过相应 ColorBlend 对象(Rainbow5,红、黄、绿、青、蓝)构造相应 VectorStyle 对象绘制要素。代码如下:

```
VectorStyle min = new VectorStyle();
VectorStyle max = new VectorStyle();
GradientTheme popdens = new GradientTheme("PopDens", 0, 400, min, max);
popdens. FillColorBlend = ColorBlend. Rainbow5;
```

```
layCountries.Theme = popdens;
```

同时为了使面积大的国家能够有比较大的标注，这里还设置了标注图层的渐变样式。这里需要先实例化两个标注样式对象，表示最小、最大样式，设置最小样式的前景色为黑色、字体为"GenericSerif"，字体大小为 6 号并且背景色与最大样式相同，都为黑色，最大样式的前景色为蓝色、字体为"GenericSerif"，字体大小为 9 号，最后将这两个样式作为参数实例化渐变主题 GradientTheme，并赋予注记图层的 Theme 属性：

```
LabelStyle lblMin = new LabelStyle();
LabelStyle lblMax = new LabelStyle();
lblMin.ForeColor = Color.Black;
lblMin.Font = new Font(FontFamily.GenericSerif, 6);
lblMax.ForeColor = Color.Blue;
lblMax.BackColor = new SolidBrush(Color.FromArgb(128, 255, 255, 255));
lblMin.BackColor = lblMax.BackColor;
lblMax.Font = new Font(FontFamily.GenericSerif, 9);
layLabel.Theme = new GradientTheme("PopDens", 0, 400, lblMin, lblMax);
```

城市图层的设置与国家图层相同，可以参照以上说明。

9.3.2.3 GdalSample 类

在使用这个类之前，先安装 FWTools 程序，然后配置 App.config 文件中的 value 属性的值为 FWTools 程序安装完成后相应的目录。代码如下：

```
<add key="FWToolsBinPath" value="C:\Program Files\FWTools2.4.7\bin"/>
<add key="FWToolsProjLib" value="C:\Program Files\FWTools2.4.7\proj_lib"/>
<add key="FWToolsGeoTiffCsv" value="C:\Program Files\FWTools2.4.7\data"/>
<add key="FWToolsGdalData" value="C:\Program Files\FWTools2.4.7\data"/>
<add key="FWToolsGdalDriver" value="C:\Program Files\FWTools2.4.7\plugins"/>
```

GdalSample 使用了 SharpMap.Extensions 名称空间下的扩展类 GdalRasterLayer，这个类实现了对 Gdal 格式数据的访问，也是 Sharpmap 的高扩展性的体现之一，有关该类的详细信息可以参照本书第 11 章，该章专门介绍了有关 GDAL 数据格式的操作。这里仅需知道，这个类实现了 IProvider 接口，使用时只需设置数据的路径，即可实现对该类数据的访问。

GdalSample 数据类提供了两类五套数据，同样有一个整型变量控制输出数据，代码如下：

```
public static Map InitializeMap()
{
    switch (_num++ % 9)
    {
        case 0:
        case 1:
        case 2:
        case 3:
        case 4:
```

```
                return InitializeGeoTiff(_num);
            default:
                return InitializeVRT(ref _num);
        }
    }
```

_num 的初始值为 0,在 InitializeMap 函数中,先对_num 自加,当自加表达式的值为 0、1、2、3、4 时,调用 InitializeGeoTiff 初始化地图,否则调用 InitializeMap 初始化。

A　InitializeGeoTiff 函数

该函数构造的是 GeoTiff 目录下的数据,通过变量_num 的值,实例化不同的地图,代码实例化地图对象后,设置地图的背景色为白色、数据文件所在的相对路径为"GeoData/GeoTiff/",代码如下:

```
Map map = new Map();
map.BackColor = Color.White;
const string relativePath = "GeoData/GeoTiff/";
```

接着声明一个 GdalRasterLayer 类型的变量,以保存通过 index 控制生成的不同的图层,并且在 GdalRasterLayer 的构造函数中设置图层的名称为"GeoTiff";

```
GdalRasterLayer layer;
switch (index)
{
    case 2:
        layer = new GdalRasterLayer("GeoTiff", relativePath + "utm.tif");
        layer.UseRotation = true;
        map.Layers.Add(layer);
        break;
    case 3:
        layer = new GdalRasterLayer("GeoTiff", relativePath + "utm.jp2");
        layer.UseRotation = true;
        map.Layers.Add(layer);
        break;
    case 4:
        layer = new GdalRasterLayer("GeoTiff", relativePath +
            "world_raster_mod.tif");
        layer.UseRotation = true;
        map.Layers.Add(layer);
        break;
    default:
        if (!File.Exists(relativePath + "format01-image_a.tif"))
        {
            throw new Exception("Make sure the data is in the relative directory: " +
                relativePath);
```

　　　　　　}
　　　}

然后加载位于相对路径中的数据，共 4 幅数据，将其加入地图对象 map 中，关键代码如下：

```
layer = new GdalRasterLayer("GeoTiffA", relativePath + "format01 - image_a.tif");
map.Layers.Add(layer);
```

然后加载位于相对路径下的边界矢量文件，设置样式为现实黑色边框、透明填充色，并且允许使用样式：

```
VectorLayer shapeLayer;
shapeLayer = new VectorLayer("outline", new ShapeFile(relativePath + "outline.shp"));
shapeLayer.Style.Fill = Brushes.Transparent;
shapeLayer.Style.Outline = Pens.Black;
shapeLayer.Style.EnableOutline = true;
shapeLayer.Style.Enabled = true;
map.Layers.Add(shapeLayer);
break;
```

最后调用 map.ZoomToExtents() 方法全视图显示地图，判断 _num 的值大于 5 时就设置其值为 1 以控制输出数据。设置 gdalSampleDataset 变量为"GeoTiff"和_num 结合，并返回地图：

```
map.ZoomToExtents();
if (_num > 5) _num = 1;
_gdalSampleDataset = "GeoTiff" + _num;
return map;
```

B　InitializeVRT() 函数

该函数比较简单，主要读取 Gdal 虚拟栅格文件。这里使用 Vrts 静态只读数组存放地图数据的文件名称，用静态常量 RelativePath 存储地图数据的相对路径。代码如下：

```
private static readonly string[] Vrts = new string[] { @"..\DEM\Golden_CO.dem", "contours_sample_polyline_play_polyline.asc", "contours_sample_polyline_play1_polyline.vrt", "contours_sample_polyline_play2_polyline.vrt", "contours_sample_polyline_play3_polyline.vrt", "contours_sample_polyline_play3_polyline.vrt" };
private const string RelativePath = "GeoData/VRT/";
```

然后根据控制变量的值来加载不同的数据，加载方式同前，代码如下：

```
GdalRasterLayer layer = new GdalRasterLayer("VirtualRasterTable", RelativePath + Vrts[ind]);
map.Layers.Add(layer);
```

9.3.2.4　OgrSample 类

使用这个类之前也需先配置 FWTools 程序，配置方法如前所述（GdalSample）。

OgrSample 数据类提供了三套数据，同样用一个整型变量控制输出数据。这里共有三个函数，分别表示不同的数据类型。

A　InitializeDXF 函数

该函数加载的是 GeoData \ SampleDXF. dxf 数据文件。先实例化一个实现了 IProvider 接口数据源对象 Ogr，参数为数据文件路径和图层编号。代码如下：

```
provider = new Ogr(d"GeoData/SampleDXF. dxf", 0);
```

最后实例化一个矢量图层对象，设置数据源为刚实例化的 provider，将其加入地图中，并全视图显示：

```
VectorLayer lay = new VectorLayer("SampleDXF", provider);
map. Layers. Add(lay);
map. ZoomToExtents();
```

B　InitializeS57 函数

该函数加载的是 GeoData \ S57 \ US5TX51M. 000 数据文件。同样先实例化一个实现了 IProvider 接口的 Ogr 对象：

```
provider = new Ogr("GeoData/S57/US5TX51M. 000");
```

对于这个数据源中的每个图层，判断是否是面状对象：

```
prov. OgrGeometryTypeString. IndexOf("Polygon") > 0
```

若是面状对象，则设置一个由随机对象（Random 类对象）rnd 生成的随机数组成的填充颜色，外边框也一样设置；否则就设置样式的线型为 rnd 生成的随机数组成的颜色。代码如下：

```
for (Int32 i = provider. NumberOfLayers - 1; i >= 0; i--)
{
    Ogr prov = new Ogr("GeoData/S57/US5TX51M. 000", i);
    if (! prov. IsFeatureDataLayer) continue;
    string name = prov. LayerName;
    System. Diagnostics. Debug. WriteLine(string. Format("Layer {0}: {1}", i, name));
    lay = new VectorLayer(string. Format("Layer_{0}", name), prov);
    if (prov. OgrGeometryTypeString. IndexOf("Polygon") > 0)
    {
        lay. Style. Fill =
            new SolidBrush(Color. FromArgb(150, Convert. ToInt32(rnd. NextDouble() *
            255), Convert. ToInt32(rnd. NextDouble() * 255),
            Convert. ToInt32(rnd. NextDouble() * 255)));
        lay. Style. Outline =
            new Pen(
                Color. FromArgb(150, Convert. ToInt32(rnd. NextDouble() * 255),
                    Convert. ToInt32(rnd. NextDouble() * 255),
```

```
                        Convert.ToInt32(rnd.NextDouble() * 255)),
            Convert.ToInt32(rnd.NextDouble() * 3));
    lay.Style.EnableOutline = true;
}
else
{
    lay.Style.Line =
        new Pen(
            Color.FromArgb(150, Convert.ToInt32(rnd.NextDouble() * 255),
                Convert.ToInt32(rnd.NextDouble() * 255),
                Convert.ToInt32(rnd.NextDouble() * 255)),
                Convert.ToInt32(rnd.NextDouble() * 3));
}
map.Layers.Add(lay);
```

C InitializeMapinfo 函数

该函数加载的是 GeoData/MapInfo 目录下的数据。实例化过程与 InitializeDXF 函数实例化图层相似，只不过需修改数据路径为相应值：

```
VectorLayer layCountries = new VectorLayer("Countries");
layCountries.DataSource = new Ogr("GeoData/MapInfo/countriesMapInfo.tab");
```

然后设置样式为绿色黑边填充、SIRD 为 4326（地图投影中最简单的坐标系统）：

```
//Set fill-style to green
layCountries.Style.Fill = new SolidBrush(Color.Green);
//Set the polygons to have a black outline
layCountries.Style.Outline = Pens.Black;
layCountries.Style.EnableOutline = true;
layCountries.SRID = 4326;
```

其余两个数据层和标注层加载类似，不再赘述。

9.3.2.5 TileLayerSample 类

TitleLayerSample 数据类使用了包含 BingMap 及 GoogleMap 提供的瓦片式数据服务等 9 种地图数据，实现了对瓦片数据服务的访问。在 InitializeMap 中同样是通过一个整形变量控制输出数据，共有四个函数。

InitializeMapOsm 函数实现对 OpenStreetMap 地图（一个可编辑的免费世界地图数据，网址为 http://www.openstreetmap.org/）的访问，这里先实例化一个地图对象。在实例化一个瓦片式数据图层对象 TileLayer（实现了 IProvider），参数为 OsmTileSource 对象和图层的名称，构造函数中的 OsmTileSource 实现了对默认地图（http://b.tile.openstreetmap.org）的访问。至此，完成了 TileLayer 图层新建工作，然后将这个新的图层加入地图对象 Map 中。代码如下：

```
Map map = new Map();
```

```
TileLayer tileLayer = new TileLayer(new OsmTileSource(), "TileLayer - OSM");
map.Layers.Add(tileLayer);
map.ZoomToBox(tileLayer.Envelope);
return map;
```

InitializeMapOsmWithXls 函数实现了对瓦片式数据和 Xls 格式的数据的访问。先实例化一个瓦片式图层，设置图层的名称为"TileLayer - OSM with XLS"，并将其加入地图对象中。代码如下：

```
Map map = new Map();
TileLayer tileLayer = new TileLayer(new OsmTileSource(), "TileLayer - OSM with XLS");
map.Layers.Add(tileLayer);
```

然后打开 Excel 数据，先设置 Xsl 数据的路径，通过 System.Data.OleDb 数据访问组件读取 .xls 数据并填充到 ds 中：

```
//Get data from excel
var xlsPath = string.Format(XlsConnectionString, System.IO.Directory.GetCurrentDirectory(),
"GeoData\\Cities.xls");
var ds = new System.Data.DataSet("XLS");
using (var cn = new System.Data.OleDb.OleDbConnection(xlsPath))
{
    cn.Open();
    using (var da = new System.Data.OleDb.OleDbDataAdapter(new
    System.Data.OleDb.OleDbCommand("SELECT * FROM [Cities$]", cn)))
        da.Fill(ds);
}
```

然后为 ds 的第一个数据表（.xls 数据）添加一个 Rotation 字段，以存储旋转角度，这里将旋转角度设置为 -angle：

```
//Add Rotation Column
ds.Tables[0].Columns.Add("Rotation", typeof(float));
foreach (System.Data.DataRow row in ds.Tables[0].Rows)
    row["Rotation"] = -angle;
```

建立数据源对象，这里使用了 SharpMap.Extensions 扩展中的 DataTablePiont 类，实现了对点对象的封装，这个实现了 IProvider 接口，将实例化后的 DataTablePiont 作为参数，新建一个矢量图层对象，设置样式的符号为矢量图层默认符号：

```
//Set up provider
var xlsProvider = new SharpMap.Data.Providers.DataTablePoint(ds.Tables[0], "OID", "X", "Y");
var xlsLayer = new SharpMap.Layers.VectorLayer("XLS", xlsProvider);
xlsLayer.Style.Symbol = SharpMap.Styles.VectorStyle.DefaultSymbol;
```

由于使用的 Excel 数据的地理坐标系是 EPSG：4326 标准定义的，需将其转换为 OSM 标准定义的地理坐标系。这里使用 Proj4 项目的投影变换，先创建一个坐标变换工厂对象，用于实现坐标变换：

  var ctf = new ProjNet. CoordinateSystems. Transformations. CoordinateTransformationFactory();

然后创建一个坐标系统工厂对象 cf，用于创建不同的坐标系统，这里调用 cf 的 CreateFromWkt 方法创建了两个坐标系统 epsg3785、epsg3785：

  var cf = new ProjNet. CoordinateSystems. CoordinateSystemFactory();
  var epsg4326 = cf. CreateFromWkt("GEOGCS[\"WGS 84 \",DATUM[\"WGS_1984 \",SPHEROID[\"WGS 84 \",6378137,298. 257223563,AUTHORITY[\"EPSG \", \"7030 \"]],AUTHORITY[\"EPSG \", \"6326 \"]],PRIMEM[\"Greenwich \",0,AUTHORITY[\"EPSG \", \"8901 \"]],UNIT[\"degree \",0. 01745329251994328,AUTHORITY[\"EPSG \", \"9122 \"]],AUTHORITY[\"EPSG \", \"4326 \"]]");
  var epsg3785 = cf. CreateFromWkt("PROJCS[\"Popular Visualisation CRS / Mercator \",GEOGCS[\"Popular Visualisation CRS \",DATUM[\"Popular Visualisation Datum \",SPHEROID[\"Popular Visualisation Sphere \",6378137,0,AUTHORITY[\"EPSG \", \"7059 \"]],TOWGS84[0,0,0,0,0,0,0],AUTHORITY[\"EPSG \", \"6055 \"]],PRIMEM[\"Greenwich \",0,AUTHORITY[\"EPSG \", \"8901 \"]],UNIT[\"degree \",0. 0174532925199433,AUTHORITY[\"EPSG \", \"9102 \"]],AXIS[\"E \",EAST],AXIS[\"N \",NORTH],AUTHORITY[\"EPSG \", \"4055 \"]],PROJECTION[\"Mercator \"],PARAMETER[\"False_Easting \",0],PARAMETER[\"False_Northing \",0],PARAMETER[\"Central_Meridian \",0],PARAMETER[\"Latitude_of_origin \",0],UNIT[\"metre \",1,AUTHORITY[\"EPSG \", \"9001 \"]],AXIS[\"East \",EAST],AXIS[\"North \",NORTH],AUTHORITY[\"EPSG \", \"3785 \"]]");

最后将图层的 CoordinateTransformation 属性设置为变换工厂执行坐标变换方法 CreateFromCoordinateSystems 后的坐标系统：

  xlsLayer. CoordinateTransformation = ctf. CreateFromCoordinateSystems(epsg4326,epsg3785);

为 .xls 数据源对象创建一个标注图层：

  var xlsLabelLayer = new SharpMap. Layers. LabelLayer("XLSLabel");
  xlsLabelLayer. DataSource = xlsProvider;
  xlsLabelLayer. LabelColumn = "Name";
  xlsLabelLayer. PriorityColumn = "Population";
  xlsLabelLayer. Style. CollisionBuffer = new System. Drawing. SizeF(2f, 2f);
  xlsLabelLayer. Style. CollisionDetection = true;
  xlsLabelLayer. LabelFilter =
  SharpMap. Rendering. LabelCollisionDetection. ThoroughCollisionDetection;
  xlsLabelLayer. CoordinateTransformation = xlsLayer. CoordinateTransformation;
  map. Layers. Add(xlsLabelLayer);

设置标注图层的数据源为前面建立的 xlsProvider，用于标注的属性列的名称为"Name"，标注的优先级依据"Population"字段，冲突缓冲区为两个像素大小 System. Drawing. SizeF(2f, 2f)，设置允许冲突检测，冲突检测的方式为压盖检测 LabelCollisionDetection. ThoroughCollisionDetection，同时设置标注图层的坐标变换属性 CoordinateTransformation 的值为

xlsLayer.CoordinateTransformation。

9.3.2.6 其他类

WmsSample 类示例了如何使用 WMS 服务所提供的空间数据；OgrSample 类展示了如何使用 OGR 数据源；PostGisSample、OracleSample、SpatiaLiteSample、WfsSample 类要安装、配置相应的空间数据库或服务才能使用，由于篇幅关系，本书不做详细说明，其原理与上面介绍的其他类的原理是一致的。

复习思考题

9-1 思考理解 WinFormSamples 示例访问多种数据源的方式。

9-2 运行 SharpMap 自带的 WinFormSamples 示例，试编程对其进行扩展，对更多的数据源进行访问。

第 10 章 Windows 应用程序开发
——DemoWinForm

10.1 数　据

本应用程序中使用的是 shape 格式的数据，系统自带了一些数据，在项目目录下的 Data 目录下。需说明的是，该应用程序提供了加载 shape 数据的功能，并不限于程序自带数据。

10.2 系统简介

DemoWinForm 示例中，实现了对 ShapeFile 格式的数据文件进行加载、显示、缩放、查询操作和随机生成点、线、面空间数据的功能，同时实现了简单的图层目录的显示和控制，系统运行界面如图 10 – 1 所示。

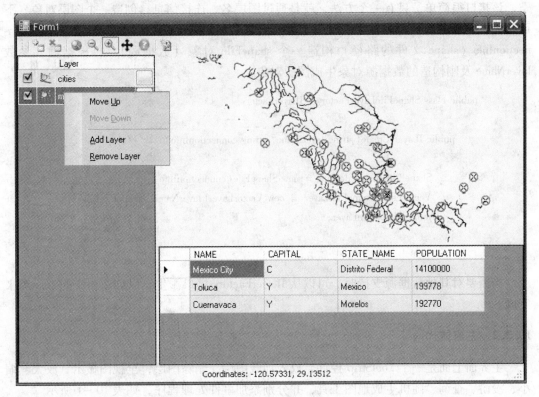

图 10 – 1　DemoWinForm 运行界面

图 10-1 所示界面中，左边采用的是 DataGridView 显示图层的信息，分别是图层的图标、图层的名称。同时这个 DataGridView 控件还绑定了上下文菜单，可以实现图层的顺序调整、加载和删除图层。

运行界面中间是一个 MapImage 控件，用于现实地图和交互操作。

运行界面上边是一个工具条，包含了加载数据、删除数据、全视图显示、放大、缩小、要素查询和随机生成数据功能。

运行界面右下方也是一个 DataGridView 控件，该控件专门用于显示要素查询的结果。

10.3 代码分析

10.3.1 数据访问代码

虽然本例中只使用了 Shape 格式数据，但为了扩展对不同格式数据访问功能，系统采用了数据工厂的设计方式，首先定义一个数据访问接口 ILayerFactory，其定义如下：

```csharp
public interface ILayerFactory
{
    ILayer Create(string layerName, string connectionInfo);
}
```

该接口很简单，只有一个方法，就是通过图层名、连接信息，创造一个图层对象。下面我们看一下访问 Shape 格式数据工厂类的实现代码。代码中，首先根据输入参数 connectionInfo（shape 文件的路径）构造一个 shapeFile 对象（数据源），再根据输入参数 layerName 及刚构造的数据源对象生成图层对象：

```csharp
public class ShapeFileLayerFactory : ILayerFactory
{
    public ILayer Create(string layerName, string connectionInfo)
    {
        ShapeFile shapeFileData = new ShapeFile(connectionInfo);
        VectorLayer shapeFileLayer = new VectorLayer(layerName, shapeFileData);
        return shapeFileLayer;
    }
}
```

当需要对其他数据源支持时，可以从 ILayerFactory 接口派生，以创建相应的工厂类来完成。

10.3.2 主窗体代码

主界面上面是一个 ToolStrip 控件，放入加载图层、删除图层、全视图显示、放大、缩小、漫游、查询、随机生成地图工具，并分别添加事件处理程序，如表 10-1 所示。

在进行这些操作的过程中，都是用了窗体的 BeginInvoke 方法，此方法执行一个异步委托。这个方法的参数的设置使用了匿名委托,将此匿名委托强制转换为 MethodInvoker 委托类型后作为参数传入。这些异步委托对应的函数如表 10-2 所示。

表 10-1 ToolStrip 控件中各个按钮功能、名称、事件处理函数对应表

功 能	名 称	事件处理函数
加载图层	AddLayerToolStripButton	AddLayerToolStripButton_Click
删除图层	RemoveLayerToolStripButton	RemoveLayerToolStripButton_Click
全视图显示	ZoomToExtentsToolStripButton	ZoomToExtentsToolStripButton_Click
放 大	ZoomInModeToolStripButton	ZoomInModeToolStripButton_Click
缩 小	ZoomOutModeToolStripButton	ZoomOutModeToolStripButton_Click
漫 游	PanToolStripButton	PanToolStripButton_Click
查 询	QueryModeToolStripButton	QueryModeToolStripButton_Click
随机生成地图	AddNewRandomGeometryLayer	AddNewRandomGeometryLayer_Click

表 10-2 ToolStrip 控件中各个按钮事件处理函数与异步委托对应表

事件处理函数	异 步 委 托
AddLayerToolStripButton_Click	BeginInvoke((MethodInvoker)delegate{ loadLayer(); })
RemoveLayerToolStripButton_Click	BeginInvoke((MethodInvoker)delegate{ removeLayer(); })
ZoomToExtentsToolStripButton_Click	BeginInvoke((MethodInvoker)delegate{ zoomToExtents(); })
ZoomInModeToolStripButton_Click	BeginInvoke((MethodInvoker)delegate{ changeMode(MapImage.Tools.ZoomIn); })
ZoomOutModeToolStripButton_Click	BeginInvoke((MethodInvoker)delegate{ changeMode(MapImage.Tools.ZoomOut); })
PanToolStripButton_Click	BeginInvoke((MethodInvoker)delegate{ changeMode(MapImage.Tools.Pan); })
QueryModeToolStripButton_Click	BeginInvoke((MethodInvoker)delegate{ changeMode(MapImage.Tools.Query); })
AddNewRandomGeometryLayer_Click	BeginInvoke((MethodInvoker)delegate{ addNewRandomGeometryLayer(); })

A 加载地图 loadLayer()

为方便对图层创建工作,使用了 ShapeFileLayerFactory 来完成图层的创建。首先根据文件的后缀名,等到相应的 LayerFactory,调用其 Create 完成图层的创建。这里的_layerFactoryCatalog 是一个 Dictionary<string, ILayerFactory>类型的对象,用于保存工厂对象,在窗体的构造函数中,调用_layerFactoryCatalog [".shp"] = new ShapeFileLayerFactory() 注册了 ShapeFile 工厂对象。最后调用 addLayer 方法加载图层,并调用 changeUIOnLayerSelectionChange 方法更新界面。主要代码如下:

```
private void loadLayer()
{
```

```
DialogResult result = AddLayerDialog.ShowDialog(this);
if (result == DialogResult.OK)
{
    foreach (string fileName in AddLayerDialog.FileNames)
    {
        string extension = Path.GetExtension(fileName);
        ILayerFactory layerFactory = null;
        if (!_layerFactoryCatalog.TryGetValue(extension, out layerFactory))
            continue;
        ILayer layer = layerFactory.Create(Path.GetFileNameWithoutExtension(fileName), fileName);
        addLayer(layer);
    }
    changeUIOnLayerSelectionChange();
    MainMapImage.Refresh();
}
```

最后就是调用添加图层方法 addLayer，将创建好的图层加入 MapImage 控件的地图容器 Map 中，并在图层目录中加入图层信息，包括图层的可见性、图层类型图标、图层名称，代码如下：

```
private void addLayer(ILayer layer)
{
    MainMapImage.Map.Layers.Add(layer);
    LayersDataGridView.Rows.Insert(0, true,
        getLayerTypeIcon(layer.GetType()), layer.LayerName);
}
```

B 删除地图 removeLayer

首先获取在图层目录中选择的要删除图层的名称，然后通过对比名称，查找需要删除的图层，代码如下：

```
string layerName = LayersDataGridView.SelectedRows[0].Cells[2].Value as string;
ILayer layerToRemove = null;
foreach (ILayer layer in MainMapImage.Map.Layers)
    if (layer.LayerName == layerName)
    {
        layerToRemove = layer;
        break;
    }
```

调用 MapImage.Map.layers.Remove 方法删除图层，并在地图目录中删除相应信息：

```
MainMapImage.Map.Layers.Remove(layerToRemove);
```

LayersDataGridView. Rows. Remove(LayersDataGridView. SelectedRows[0]);

 C 全视图显示 zoomToExtents ( )

直接调用用 MapImage. Map. ZoomToExtent 方法实现全视图显示，并刷新地图。代码如下：

MainMapImage. Map. ZoomToExtents( );
MainMapImage. Refresh( );

 D 缩小、放大、漫游、查询

地图操作工具放大、缩小、漫游、查询等，只需要更改工具即可用 MapImage 封装好的事件处理函数完成相应操作。代码如下：

```
private void changeMode( MapImage. Tools tool)
{
    MainMapImage. ActiveTool = tool;
    ZoomInModeToolStripButton. Checked = (tool = = MapImage. Tools. ZoomIn);
    ZoomOutModeToolStripButton. Checked = (tool = = MapImage. Tools. ZoomOut);
    PanToolStripButton. Checked = (tool = = MapImage. Tools. Pan);
    QueryModeToolStripButton. Checked = (tool = = MapImage. Tools. Query);
}
```

为实现查询功能，窗体的构造函数为 ImageMap 控件添加了 MapQueried 事件：

MainMapImage. MapQueried + = MainMapImage_MapQueried;

MainMapImage_MapQueried 函数使用查询后的要素数据表作为 FeatureDataGridView 的数据源，以显示查询的结果：

```
private void MainMapImage_MapQueried( FeatureDataTable data)
{
    FeaturesDataGridView. DataSource = data;
}
```

 E 随机生成地图

系统随机生成点、线、面数据，点数据使用预先放置在资源文件 Resources 中的图片作为符号（可通过窗体的 Properties. Resources 属性访问，具体访问原理请参照相关 .NET 书籍），这些符号保存在主窗体的只读成员变量 _symbolTable 中，这个变量在窗体的构造函数中，通过 registerSymbols 方法初始化：

private readonly Dictionary < string, Bitmap > _symbolTable = new Dictionary < string, Bitmap > ( );

初始化的过程就是访问资源文件里存储的图片，共保存 7 张 png 格式的图片，代码如下：

```
private void registerSymbols( )
{
    _symbolTable["Notices"] = Resources. Chat;
    _symbolTable["Radioactive Fuel Rods"] = Resources. DATABASE;
```

```csharp
_symbolTable["Bases"] = Resources.Flag;
_symbolTable["Houses"] = Resources.Home;
_symbolTable["Measures"] = Resources.PIE_DIAGRAM;
_symbolTable["Contacts"] = Resources.Women;
_symbolTable["Prospects"] = Resources.Women_1;
}
```

为了得到随机的填充颜色，这里将系统已知的颜色进行遍历，存储在主窗体的只读成员变量_colorTable 中；_colorTable 的初始化也是在构造函数中通过调用 registerKnownColors 方法完成初始化，这里遍历 System.Drawing.knownColor 中定义的已知颜色的名称和值存储在_colorTable 中：

```csharp
private readonly Dictionary<string, Color> _colorTable = new Dictionary<string, Color>();
private static void registerKnownColors(Dictionary<string, Color> colorTable)
{
    foreach (string colorName in Enum.GetNames(typeof(KnownColor)))
    {
        KnownColor color = (KnownColor)Enum.Parse(typeof(KnownColor), colorName);
        colorTable[colorName] = Color.FromKnownColor(color);
    }
}
```

接下来就是随机生成图层数据，addNewRandomGeometryLayer 就是完成这项任务的。随机调用 generatePoints、generateLines、generatePolygons 函数之一分别产生点、线、面数据，再从之前配置好的 symbolEntry、colorEntry 容器中随机取得相应对象作为点或线、面的符号（颜色）；最后构造图层，并加入到地图对象中，代码如下：

```csharp
private void addNewRandomGeometryLayer()
{
    Random rndGen = new Random();
    Collection<Geometry> geometry = new Collection<Geometry>();

    VectorLayer layer = new VectorLayer(String.Empty);

    switch (rndGen.Next(3))
    {
        case 0:
        {
            generatePoints(geometry, rndGen);
            KeyValuePair<string, Bitmap> symbolEntry = getSymbolEntry(rndGen.Next(_symbolTable.Count));
            layer.Style.Symbol = symbolEntry.Value;
            layer.LayerName = symbolEntry.Key;
        }
```

```
            break;
        case 1:
        {
            generateLines(geometry, rndGen);
            KeyValuePair<string, Color> colorEntry = getColorEntry(rndGen.Next(_color-
Table.Count));
            layer.Style.Line = new Pen(colorEntry.Value);
            layer.LayerName = String.Format("{0} lines", colorEntry.Key);
        }
            break;
        case 2:
        {
            generatePolygons(geometry, rndGen);
            KeyValuePair<string, Color> colorEntry = getColorEntry(rndGen.Next(_color-
Table.Count));
            layer.Style.Fill = new SolidBrush(colorEntry.Value);
            layer.LayerName = String.Format("{0} squares", colorEntry.Key);
        }
            break;
        default:
            throw new NotSupportedException();
    }

    GeometryProvider provider = new GeometryProvider(geometry);
    layer.DataSource = provider;

    addLayer(layer);
}
```

generatePoints 函数随机生成点数据,点数在 10~100 之间,点的坐标在 0~1000 之间:

```
private void generatePoints(Collection<Geometry> geometry, Random rndGen)
{
    int numPoints = rndGen.Next(10, 100);
    for (int pointIndex = 0; pointIndex < numPoints; pointIndex++)
    {
        GeoPoint point = new GeoPoint(rndGen.NextDouble() * 1000, rndGen.NextDouble() *
1000);
        geometry.Add(point);
    }
}
```

generateLines 函数随机生成线数据,线数在 10~100 之间,线的坐标在 0~1000 之间:

```csharp
private void generateLines(Collection<Geometry> geometry, Random rndGen)
{
    int numLines = rndGen.Next(10, 100);
    for (int lineIndex = 0; lineIndex < numLines; lineIndex++)
    {
        LineString line = new LineString();
        Collection<GeoPoint> verticies = new Collection<GeoPoint>();

        int numVerticies = rndGen.Next(4, 15);

        GeoPoint lastPoint = new GeoPoint(rndGen.NextDouble() * 1000, rndGen.NextDouble() * 1000);
        verticies.Add(lastPoint);

        for (int vertexIndex = 0; vertexIndex < numVerticies; vertexIndex++)
        {
            GeoPoint nextPoint = new GeoPoint(lastPoint.X + rndGen.Next(-50, 50),
                                              lastPoint.Y + rndGen.Next(-50, 50));
            verticies.Add(nextPoint);

            lastPoint = nextPoint;
        }

        line.Vertices = verticies;

        geometry.Add(line);
    }
}
```

generatePolygons 函数随机生成面数据，面数在 10～100 之间，面的坐标在 0～1000 之间，每个面由 4 个点构成：

```csharp
private void generatePolygons(Collection<Geometry> geometry, Random rndGen)
{
    int numPolygons = rndGen.Next(10, 100);
    for (int polyIndex = 0; polyIndex < numPolygons; polyIndex++)
    {
        Polygon polygon = new Polygon();
        Collection<GeoPoint> verticies = new Collection<GeoPoint>();
        GeoPoint upperLeft = new GeoPoint(rndGen.NextDouble() * 1000, rndGen.NextDouble() * 1000);
        double sideLength = rndGen.NextDouble() * 50;

        // Make a square
```

```
            verticies. Add( upperLeft );
            verticies. Add( new GeoPoint( upperLeft. X + sideLength, upperLeft. Y));
            verticies. Add( new GeoPoint( upperLeft. X + sideLength, upperLeft. Y - sideLength));
            verticies. Add( new GeoPoint( upperLeft. X, upperLeft. Y - sideLength));
            polygon. ExteriorRing = new LinearRing( verticies );

            geometry. Add( polygon );
        }
    }
```

F 实现坐标实时显示鼠标的位置

主窗体显示地图的控件 MapImage，名称为 MainMapImage，项目中设置 MoseMove 事件处理函数为 MainMapImage_MouseMove，实现在鼠标移动时实时显示鼠标的位置，这里将显示坐标控件的文本属性设置为坐标的 X、Y 值：

```
private void MainMapImage_MouseMove( GeoPoint WorldPos, MouseEventArgs ImagePos)
{
    CoordinatesLabel. Text = String. Format("Coordinates: {0:N5}, {1:N5}",
    WorldPos. X, WorldPos. Y);
}
```

复习思考题

10-1 运行 SharpMap 自带的 DemoWinForm 示例，理解随机生成数据功能模块，试画出算法流程图。

第 11 章 数据源扩展与图层对象扩展

系统的可扩展性是衡量系统优劣的一项重要标准。SharpMap 的系统设计具有良好的可扩展性，SharpMap 的 Extensions 项目中有关于 Data（数据）及 Layers（图层）的一些扩展对象。本章将就这两种扩展类型进行分析说明。

当前 GIS 系统中，存在着各种不同的 GIS 数据格式，因此对数据源的访问提供扩展功能是非常必要的。

11.1 DataTablePoint 类

DataTablePoint 类提供点数据源的功能，使用 System.Data.DataTable 作为数据存储对象，数据表中至少包含两列 double 类型的数据，表示点的 X、Y 坐标值及一个整型的数据列，整型的值是唯一的，用于标识每个数据行，数据表可以有额外的属性数据列，类似于如表 11－1 所示格式的数据。

表 11－1 数据属性

OID	NAME	POPULATION	CAPITAL	X	Y
1	Petropavloski-Kamchatskiy	269000	N	158.7200012	53.20000076
2	Sapporo	1900000	N	141.3454742	43.05525208
3	Aomori	294045	N	140.6699982	40.75

在示例程序 WinFormSamples 中，TileLayerSample 类中使用该类访问存储在 Excel 文件中的点数据。

11.1.1 DataTablePoint 类的数据成员

disposed 指示数据源是已经执行回收，代码如下：

```
private bool disposed = false;
```

11.1.2 DataTablePoint 类的属性

11.1.2.1 ConnectionID

ConnectionID 表示连接数据源的连接字符串，是只读属性，代码如下：

```
private string _ConnectionString;
public string ConnectionID
{
    get { return _ConnectionString; }
}
```

11.1.2.2 IsOpen

IsOpen 表示数据源是否打开,是只读属性,代码如下:

```
private bool _IsOpen;
public bool IsOpen
{
    get { return _IsOpen; }
}
```

11.1.2.3 SRID

获取和设置空间参考编号,代码如下:

```
private int _SRID = -1;
public int SRID
{
    get { return _SRID; }
    set { _SRID = value; }
}
```

11.1.2.4 Table

设置和获取数据表对象,代码如下:

```
private DataTable _Table;
public DataTable Table
{
    get { return _Table; }
    set { _Table = value; }
}
```

11.1.2.5 ObjectIdColumn

设置和获取存对象 ID 编号的列的名称,代码如下:

```
private string _ObjectIdColumn;
public string ObjectIdColumn
{
    get { return _ObjectIdColumn; }
    set { _ObjectIdColumn = value; }
}
```

11.1.2.6 XColumn

设置和获取存储点 X 坐标值的列的名称,代码如下:

```
private string _XColumn;
public string XColumn
{
```

```
        get { return _XColumn; }
        set { _XColumn = value; }
}
```

11.1.2.7 YColumn

设置和获取存储点 Y 坐标值的列的名称，代码如下：

```
private string _YColumn;
public string YColumn
{
        get { return _YColumn; }
        set { _YColumn = value; }
}
```

11.1.2.8 ConnectionString

设置和获取连接数据源的连接字符串，代码如下：

```
public string ConnectionString
{
        get { return _ConnectionString; }
        set { _ConnectionString = value; }
}
```

11.1.2.9 DefinitionQuery

设置和获取自定义查询字符串，代码如下：

```
private string _defintionQuery;
public string DefinitionQuery
{
        get { return _defintionQuery; }
        set { _defintionQuery = value; }
}
```

11.1.3 DataTablePoint 类的方法

11.1.3.1 DataTablePoint

构造函数，需传入数据表、表中存储对象 ID 编号列的名称、X 坐标值的列的名称、Y 坐标值的列的名称。构造函数中，保存着几个参数的值到相应的数据成员，代码如下：

```
public DataTablePoint( DataTable dataTable, string oidColumnName,
                       string xColumn, string yColumn)
{
        Table = dataTable;
        XColumn = xColumn;
        YColumn = yColumn;
        ObjectIdColumn = oidColumnName;
```

11.1.3.2 ~DataTablePoint

此为析构函数。这里 Dispose() 方法完成资源回收，代码如下：

```
~DataTablePoint()
{
    Dispose();
}
```

11.1.3.3 Close

关闭数据源。这里设置 _IsOpen 的值为 false，代码如下：

```
public void Close()
{
    _IsOpen = false;
}
```

11.1.3.4 Dispose()

完成资源的清理。这里调用 Dispose(true) 方法检查资源是否被回收，然后调用垃圾回收器的 GC.SuppressFinalize(this) 方法通知系统该对象已经回收，不需要再进行垃圾回收，代码如下：

```
public void Dispose()
{
    Dispose(true);
    GC.SuppressFinalize(this);
}
```

11.1.3.5 Dispose(bool disposing)

设置 disposed 值，指示资源是否已被释放，代码如下：

```
internal void Dispose(bool disposing)
{
    if(!disposed)
    {
        if(disposing)
        {
        }
        disposed = true;
    }
}
```

11.1.3.6 Open

打开数据源。这里设置 _IsOpen 的值为 true，代码如下：

```csharp
public void Open()
{
    _IsOpen = true;
}
```

11.1.3.7　GetGeometriesInView

获取指定范围内的几何对象。这里先构造查询语句，调用 Table 的 Select 方法将查询的结果保存在 DataRow [ ] drow 中，然后遍历 drow 数据行数组，生成点对象并加入几何对象集合中并返回，代码如下：

```csharp
public Collection<Geometry> GetGeometriesInView(BoundingBox bbox)
{
    DataRow[] drow;
    Collection<Geometry> features = new Collection<Geometry>();
    if (Table.Rows.Count == 0)
    {
        return null;
    }
    string strSQL = XColumn + " > " + bbox.Left.ToString(Map.NumberFormatEnUs) +
        " AND " + XColumn + " < " + bbox.Right.ToString(Map.NumberFormatEnUs) +
        " AND " + YColumn + " > " + bbox.Bottom.ToString(Map.NumberFormatEnUs) +
        " AND " + YColumn + " < " + bbox.Top.ToString(Map.NumberFormatEnUs);
    drow = Table.Select(strSQL);
    foreach (DataRow dr in drow)
    {
        features.Add(new Point((double)dr[XColumn], (double)dr[YColumn]));
    }
    return features;
}
```

11.1.3.8　GetObjectIDsInView

与 GetGeometriesInView 方法相似，只不过这里返回的是在指定范围内的几何对象的 ID 编号，所以这里得到查询结果后，不用生成点对象，而是直接将对象 ID 编号值返回，代码如下：

```csharp
public Collection<uint> GetObjectIDsInView(BoundingBox bbox)
{
    DataRow[] drow;
    Collection<uint> objectlist = new Collection<uint>();
    if (Table.Rows.Count == 0)
    {
        return null;
    }
```

11.1 DataTablePoint 类

```
string strSQL = XColumn + " > " + bbox.Left.ToString(Map.NumberFormatEnUs) +
    " AND " + XColumn + " < " + bbox.Right.ToString(Map.NumberFormatEnUs) +
    " AND " + YColumn + " > " + bbox.Bottom.ToString(Map.NumberFormatEnUs) +
    + " AND " + YColumn + " < " + bbox.Top.ToString(Map, NumberFormatEnUs);
drow = Table.Select(strSQL);
foreach (DataRow dr in drow)
{
    objectlist.Add((uint)(int)dr[0]);
}
return objectlist;
}
```

11.1.3.9 GetGeometryByID

通过对象的 ID 编号获取集合对象。这里也先生成查询语句，调用 Table 的 Select 方法将查询的结果保存在 DataRow [ ] rows 中，然后对于数组中的每一行，生成几何对象并返回，所以这里返回的是崭新的对象，代码如下：

```
public Geometry GetGeometryByID(uint oid)
{
    DataRow[] rows;
    Geometry geom = null;
    if (Table.Rows.Count == 0)
    {
        return null;
    }
    string selectStatement = ObjectIdColumn + " = " + oid;
    rows = Table.Select(selectStatement);
    foreach (DataRow dr in rows)
    {
        geom = new Point((double)dr[XColumn], (double)dr[YColumn]);
    }
    return geom;
}
```

11.1.3.10 ExecuteIntersectionQuery (Geometry geom, FeatureDataSet ds)

执行相交查询，返回几何对象内的所有要素。由于点对象和其他对象相交无实际意义，这里就直接抛出异常，代码如下：

```
public void ExecuteIntersectionQuery(Geometry geom, FeatureDataSet ds)
{
    throw new NotSupportedException("ExecuteIntersectionQuery(Geometry) is not
    supported by the DataTablePoint.");
}
```

11.1.3.11 ExecuteIntersectionQuery（BoundingBox bbox，FeatureDataSet ds）

执行相交查询，返回指定外包框内的所有要素。这里先构造查询字符串，限制点的坐标值必须在指定的外包框中，调用 Table 的 Select 方法将查询的结果保存在 DataRow [ ] rows 中；然后，构建一个要素数据表对象 fdt，对 rows 中的数据行，生成相应点对象，并加入要素数据表对象 fdt 中，最后将要素数据表，加入传入的参数 ds（要素数据集），代码如下：

```
public void ExecuteIntersectionQuery(BoundingBox bbox, FeatureDataSet ds)
{
    DataRow[ ] rows;
    if (Table.Rows.Count = = 0)
    {
        return;
    }
    string statement = XColumn + " > " + bbox.Left.ToString(Map.NumberFormatEnUs)
        + " AND " + XColumn + " < " + bbox.Right.ToString(Map.NumberFormatEnUs)
        + " AND " + YColumn + " > " + bbox.Bottom.ToString(Map.NumberFormatEnUs)
        + " AND " + YColumn + " < " + bbox.Top.ToString(Map.NumberFormatEnUs);
    rows = Table.Select(statement);
    FeatureDataTable fdt = new FeatureDataTable(Table);
    foreach (DataColumn col in Table.Columns)
    {
        fdt.Columns.Add(col.ColumnName, col.DataType, col.Expression);
    }
    foreach (DataRow dr in rows)
    {
        fdt.ImportRow(dr);
        FeatureDataRow fdr = fdt.Rows[fdt.Rows.Count - 1] as FeatureDataRow;
        fdr.Geometry = new Point((double) dr[XColumn], (double) dr[YColumn]);
    }
    ds.Tables.Add(fdt);
}
```

11.1.3.12 GetFeatureCount

获取要素的个数，要素的个数即数据表的行的个数。这里直接返回行的个数，代码如下：

```
public int GetFeatureCount()
{
    return Table.Rows.Count;
}
```

11.1.3.13 GetFeature

通过行的编号获取要素数据行。这里并未实现,代码如下:

```
public FeatureDataRow GetFeature(uint RowID)
{
    throw new NotSupportedException();
}
```

11.1.3.14 GetExtents

获取数据源的外包框,这里遍历数据表中的每行,求出 X、Y 的最大最小值,然后利用最大最小值生成外包框矩形,代码如下:

```
public BoundingBox GetExtents()
{
    if (Table.Rows.Count == 0)
    {
        return null;
    }
    BoundingBox box;
    double minX = Double.PositiveInfinity,
           minY = Double.PositiveInfinity,
           maxX = Double.NegativeInfinity,
           maxY = Double.NegativeInfinity;
    foreach (DataRowView dr in Table.DefaultView)
    {
        if (minX > (double)dr[XColumn]) minX = (double)dr[XColumn];
        if (maxX < (double)dr[XColumn]) maxX = (double)dr[XColumn];
        if (minY > (double)dr[YColumn]) minY = (double)dr[YColumn];
        if (maxY < (double)dr[YColumn]) maxY = (double)dr[YColumn];
    }
    box = new BoundingBox(minX, minY, maxX, maxY);
    return box;
}
```

11.2 OgrProvider 类

GDAL/OGR 是非常著名的开源 GIS 库,GDAL 库对栅格数据进行操作,OGR 库对矢量数据进行操作,它们相当于一个通用数据访问库,甚至 ESRI 的产品中都用了此库,现在,GDAL/OGR 也提供了对 .NET 的编译支持,具体的编译与使用过程读者可自行查阅相关资料。

OgrProvider 是对 GDAL/OGR 进行包装,为 SharpMap 提供对矢量数据的访问,使用前需添加 GDAL/OGR 类库的引用。为方便使用，这里使用了名称空间的别名，解释参照注释，代码如下：

```
using Geometry = SharpMap.Geometries.Geometry;  //SharpMap 的几何对象
using OgrOgr = OSGeo.OGR.Ogr;                   //
using OgrDataSource = OSGeo.OGR.DataSource;     //Ogr 数据源
using OgrLayer = OSGeo.OGR.Layer;               //Ogr 图层
using OgrGeometry = OSGeo.OGR.Geometry;         //Ogr 几何对象
using OgrEnvelope = OSGeo.OGR.Envelope;         //Ogr 外包框矩形
using OgrFeature = OSGeo.OGR.Feature;           //Ogr 要素对象
using OgrFeatureDefn = OSGeo.OGR.FeatureDefn;   //表示要素每个属性字段的信息，//Feature-
Defn 是 一个和当前层相关的对象，它包含了该层中定义的所有属性字段
using OgrFieldDefn = OSGeo.OGR.FieldDefn;       //表述属性字段的信息
using OgrFieldType = OSGeo.OGR.FieldType;       //属性的类型
using OsrSpatialReference = OSGeo.OSR.SpatialReference;  //空间参考
using OgrGeometryType = OSGeo.OGR.wkbGeometryType;       //二进制几何类型
```

OgrProvider 类在类级别上使用了［Serializable］特性，这样就可以方便地对 OgrProvider 对象进行存储。在示例程序 WinFormSamples 项目中使用了 OgrProvider 类，OgrSample 类中使用该类访问 OGR 数据。需注意的是，若要使用本类，需安装配置 TWTools 程序，具体的配置过程参照开发应用部分的示例程序。

11.2.1 OgrProvider 类的数据成员

11.2.1.1 _bbox
表示数据源的外包框矩形。由于外包框矩形可以在使用时计算得出，这里就使用了［NonSerialized］特性，表示在序列化的过程中，该字段并不被保存，代码如下：

```
[NonSerialized]
private BoundingBox _bbox;
```

11.2.1.2 _ogrDataSource
表示 Ogr 的数据源，Datasources 可以是文件、RDBMSes、目录中的所有文件或者是被连接到本机的 Web 地址。通常，datasource 都对应一个字符串名称。这里也用［NonSerialized］特性标记，代码如下：

```
[NonSerialized]
private readonly OgrDataSource _ogrDataSource;
```

11.2.1.3 _ogrLayer
表示 Ogr 图层，这里也标记为［NonSerialized］，代码如下：

```
[NonSerialized]
private OgrLayer _ogrLayer;
```

11.2.1.4 _disposed

表示 OgrPrivider 是否已经被销毁，代码如下：

```csharp
private bool _disposed;
```

11.2.2 OgrProvider 类的属性

11.2.2.1 ConnectionID

表示数据源的连接字符串，代码如下：

```csharp
public string ConnectionID
{
    get
    {
        return string.Format("Data Source = {0};Layer{1}", _ogrDataSource.name,
            _ogrLayer.GetName());
    }
}
```

11.2.2.2 IsOpen

获取数据源的状态。当数据源状态为打开时，值为 true，否则为 false，代码如下：

```csharp
private bool _isOpen;
public bool IsOpen
{
    get { return _isOpen; }
}
```

11.2.2.3 FWToolsVersion

表示 FWTools 程序的版本，是一个只读属性，通过 FwToolsHelper 帮助类的（存储了 FwTools 程序的配置信息）FwToolsVersion 属性获取，代码如下：

```csharp
public static string FWToolsVersion
{
    get { return FwToolsHelper.FwToolsVersion; }
}
```

11.2.2.4 Filename

获取和设置数据源文件的名称，代码如下：

```csharp
private String _filename;
public string Filename
{
    get { return _filename; }
    set { _filename = value; }
}
```

11.2.2.5 DefinitionQuery

获取和设置自定义查询字符串,代码如下:

```
private String _definitionQuery = "";
public String DefinitionQuery
{
    get { return _definitionQuery; }
    set { _definitionQuery = value; }
}
```

11.2.2.6 NumberOfLayers

获取数据源图层的个数。这里使用了数据源的 GetLayerCount 方法返回图层的数目,代码如下:

```
public Int32 NumberOfLayers
{
    get
    {
        Int32 numberOfLayers = 0;
        numberOfLayers = _ogrDataSource.GetLayerCount();
        return numberOfLayers;
    }
}
```

11.2.2.7 IsFeatureDataLayer

判断图层是否为要素数据层。首先调用 Ogr 图层对象的 ResetReading() 方法,重新设置读取操作(即定位到开头),然后调用图层对象的 GetNextFeature() 方法依次返回下一个几何对象,如果返回 NULL 的话,表示全部读完。通过调用几何对象的 GetGeometryRef 方法,可以获取几何对象的指针,然后判断几何对象的类型是否为已知的要素数据类型,若不是则返回 false,代码如下:

```
public Boolean IsFeatureDataLayer
{
    get
    {
        _ogrLayer.ResetReading();
        Int32 numFeatures = _ogrLayer.GetFeatureCount(1);
        if (numFeatures <= 0) return false;
        OgrFeature feature = _ogrLayer.GetNextFeature();
        if (feature == null) return false;
        OgrGeometry geom = feature.GetGeometryRef();
        if (geom == null) return false;
        return geom.GetGeometryType() != OgrGeometryType.wkbNone;
    }
}
```

11.2.2.8　OgrGeometryTypeString

获取几何对象的类型名称。同上，取得几何对象的引用后，调用几何对象的 GetGeometryType 方法返回类型的名称，代码如下：

```
public String OgrGeometryTypeString
{
    get
    {
        _ogrLayer.ResetReading();
        Int32 numFeatures = _ogrLayer.GetFeatureCount(1);
        if (numFeatures <= 0) return string.Format("{0}",
        OgrGeometryType.wkbNone);
        OgrFeature feature = _ogrLayer.GetNextFeature();
        if (feature == null) return string.Format("{0}", OgrGeometryType.wkbNone);
        OgrGeometry geom = feature.GetGeometryRef();
        if (geom == null) return string.Format("{0}", OgrGeometryType.wkbNone);
        return string.Format("{0}", geom.GetGeometryType());
    }
}
```

11.2.2.9　LayerName

获取或设置图层的名称。获取时，调用图层的 GetLayerDefn() 方法得到图层信息，并调用 GetName() 方法得到图层名称；设置时，调用数据源的 GetLayerByName 方法获取所要设置的图层，然后赋予 _ogrLayer 变量，若不存在所要设置的图层，_ogrLayer 值不会变化，代码如下：

```
public string LayerName
{
    get { return _ogrLayer.GetLayerDefn().GetName(); }
    set
    { try
        {
            OgrLayer layer = _ogrDataSource.GetLayerByName(value);
            _ogrLayer = layer;
        }
        catch { }
    }
}
```

11.2.2.10　LayerIndex

设置和获取图层的索引号。获取时，先获取当前图层的名称，然后遍历数据源中的每个图层，若图层的名称和刚获取的图层名称相同，则返回索引号。设置时先判断索引是否超出范围，若没超出范围则调用 GetLayerByIndex 方法获取图层并赋予 _ogrLayer，代码如下：

```
public Int32 LayerIndex
{
    get
    {
        string layerName = LayerName;
        for (int i = 0; i < _ogrDataSource.GetLayerCount(); i++)
        {
            if (_ogrDataSource.GetLayerByIndex(i).GetName() == layerName)
                return i;
        }
        throw new Exception("Somehow the layer set cannot be found in datasource");
    }
    set
    {
        if (value < 0 || _ogrDataSource.GetLayerCount() - 1 < value)
            throw new ArgumentOutOfRangeException("value");
        _ogrLayer = _ogrDataSource.GetLayerByIndex(value);
    }
}
```

11.2.2.11 SRID

设置和获取空间参考的 ID 编号，代码如下：

```
private int _srid = -1;
public int SRID
{
    get { return _srid; }
    set { _srid = value; }
}
```

11.2.3 OgrProvider 类的方法

在 OgrProvider 类中有些方法使用了 ObsoleteAttribute 的特性，表示这些方法已经过时，不建议再使用，该元素在产品的未来版本中将被移除。

11.2.3.1 Ogr()

静态构造函数。该函数在类第一次被调用时执行初始化工作，调用 FwToolsHelper 辅助类配置 FWTools 程序信息，调用 OgrOgr.RegisterAll() 方法注册所有格式的驱动，代码如下：

```
static Ogr()
{
```

```
FwToolsHelper.Configure();
OgrOgr.RegisterAll();
}
```

11.2.3.2　Ogr（string filename, string layerName）

构造函数。这需传入数据源的名称 filename、图层名称 layerName，构造函数中调用 OgrOgr.Open 方法打开数据源，然后调用数据源的 GetLayerByName 方法获取图层，设置 _ogrLayer；然后调用图层的 GetSpatialRef 方法获取图层的空间参考，若空间参考不为空，调用空间参考（OsrSpatialReference）对象的 AutoIdentifyEPSG() 获取对应编号，并设置 _srid，代码如下：

```
[Obsolete("This constructor does not work well with VB.NET. Use LayerName property instead")]
public Ogr(string filename, string layerName)
{
    Filename = filename;
    _ogrDataSource = OgrOgr.Open(filename, 1);
    _ogrLayer = _ogrDataSource.GetLayerByName(layerName);
    OsrSpatialReference spatialReference = _ogrLayer.GetSpatialRef();
    if (spatialReference != null)
        _srid = spatialReference.AutoIdentifyEPSG();
}
```

11.2.3.3　Ogr（string filename, int layerNum）

构造函数。这里通过数据源的名称 filename、图层的索引号 layerNum 完成初始化，与前面的构造函数类似，不过这里是通过数据源的 GetLayerByIndex 方法获取图层，代码如下：

```
public Ogr(string filename, int layerNum)
{
    Filename = filename;
    _ogrDataSource = OgrOgr.Open(filename, 0);
    _ogrLayer = _ogrDataSource.GetLayerByIndex(layerNum);
    OsrSpatialReference spatialReference = _ogrLayer.GetSpatialRef();
    if (spatialReference != null)
        _srid = spatialReference.AutoIdentifyEPSG();
}
```

11.2.3.4　Ogr（string datasource）

构造函数。这里默认打开索引号为 0 的图层，代码如下：

```
public Ogr(string datasource)
    : this(datasource, 0)
{
}
```

11.2.3.5 Ogr(string datasource, int layerNum, out string name)

构造函数。这里调用其他构造函数后，调用图层对象的 GetName 方法，获取图层的名称，并设置 name，代码如下：

```
[Obsolete("This constructor does not work well with VB. NET. Use LayerName property instead")]
public Ogr(string datasource, int layerNum, out string name)
    : this(datasource, layerNum)
{
    name = _ogrLayer.GetName();
}
```

11.2.3.6 Ogr(string datasource, out string name)

构造函数，这里设置打开的图层的索引为 0，调用其他构造函数完成初始化，代码如下：

```
[Obsolete("This constructor does not work well with VB. NET. Use LayerName property instead")]
public Ogr(string datasource, out string name)
    : this(datasource, 0, out name)
{
}
```

11.2.3.7 ~Ogr

析构函数，这里调用 Close() 和 Dispose() 完成资源的回收，代码如下：

```
~Ogr()
{
    Close();
    Dispose();
}
```

11.2.3.8 Dispose

设置 _disposed，标识资源是否已被回收，代码如下：

```
public void Dispose()
{
    Dispose(true);
    GC.SuppressFinalize(this);
}
```

11.2.3.9 Dispose

若资源未被回收（!_disposed），就判断数据源是否为空，不为空则调用数据源的 Dispose 方法回收资源，代码如下：

```
internal void Dispose(bool disposing)
```

```
        {
            if (!_disposed)
            {
                if (disposing && _ogrDataSource != null)
                {
                    _ogrDataSource.Dispose();
                }
                _disposed = true;
            }
        }
```

11.2.3.10 GetExtents

获取当前图层的外包框矩形。这里调用图层对象的 GetExtent 方法，获取 ogr 格式的外包框，然后生成 SharpMap 对应的外包框矩形，代码如下：

```
public BoundingBox GetExtents()
{
    if (_bbox == null)
    {
        OgrEnvelope ogrEnvelope = new OgrEnvelope();
        if (_ogrLayer != null) _ogrLayer.GetExtent(ogrEnvelope, 1);
        _bbox = new BoundingBox(ogrEnvelope.MinX,
                                ogrEnvelope.MinY,
                                ogrEnvelope.MaxX,
                                ogrEnvelope.MaxY);
    }
    return _bbox;
}
```

11.2.3.11 GetFeatureCount

获取要素的个数。这里调用图层对象的 GetFeatureCount 方法返回个数，代码如下：

```
public int GetFeatureCount()
{
    return _ogrLayer.GetFeatureCount(1);
}
```

11.2.3.12 GetFeature

通过要素的行编号获取要素数据行。先构造一个要素数据集对象 fdt，调用图层的 ResetReading 方法重新设置读取操作，然后调用 ReadColumnDefinition 方法设置要素数据集的列信息，参数为要素数据集对象和图层对象，借着调用图层的 GetFeature 方法获取指定编号的要素对象，存储在 OgrFeature 类型的变量 feature 中，最后调用 OgrFeatureToFeatureDataRow 方法，将要素转化为要素数据行对象并返回此要素数据行，代码如下：

```
public FeatureDataRow GetFeature(uint rowId)
```

```
        FeatureDataTable fdt = new FeatureDataTable();
        _ogrLayer.ResetReading();
        ReadColumnDefinition(fdt, _ogrLayer);
        OgrFeature feature = _ogrLayer.GetFeature((int)rowId);
        return OgrFeatureToFeatureDataRow(fdt, feature);
    }
```

11.2.3.13 Open

打开数据源,设置_isOpen 值为 true,代码如下:

```
public void Open()
{
    _isOpen = true;
}
```

11.2.3.14 Close

关闭数据源,设置_isOpen 值为 false,代码如下:

```
public void Close()
{
    _isOpen = false;
}
```

11.2.3.15 IsOpen

返回数据源的状态,代码如下:

```
public bool IsOpen
{
    get { return _isOpen; }
}
```

11.2.3.16 GetObjectIDsInView

返回指定范围内的对象的 ID 编号集合。首先调用图层的 SetSpatialFilterRect 方法设置过滤范围,重置读取操作以重新查询,新建一个 Collection<uint> 类型的对象 objectIDs 用于存放查询结果,然后调用图层 GetNextFeature 方法遍历查询结果,调用要素的 GetFID 方法获取编号,由于图层对象 GetNextFeature 方法返回的是要素对象的拷贝,所以需要调用要素对象的 Dispose 方法,以销毁对象,代码如下:

```
public Collection<uint> GetObjectIDsInView(BoundingBox bbox)
{
    _ogrLayer.SetSpatialFilterRect(bbox.Min.X, bbox.Min.Y, bbox.Max.X, bbox.Max.Y);
    _ogrLayer.ResetReading();
    Collection<uint> objectIDs = new Collection<uint>();
    OgrFeature ogrFeature = null;
```

```csharp
        while((ogrFeature = _ogrLayer.GetNextFeature())! = null)
        {
            objectIDs.Add((uint)ogrFeature.GetFID());
            ogrFeature.Dispose();
        }
        return objectIDs;
    }
```

11.2.3.17　GetGeometryByID

通过 Ogr 形式的几何对象的 ID 编号获取对应的 SharpMap 形式的几何对象，先调用图层的 GetFeature 方法获取要素对象，然后调用 ParseOgrGeometry 方法进行转换，关于 ParseOgrGeometry 方法的详细信息，请参看后续方法解释，代码如下：

```csharp
public Geometry GetGeometryByID(uint oid)
{
    using(OgrFeature ogrFeature = _ogrLayer.GetFeature((int)oid))
        return ParseOgrGeometry(ogrFeature.GetGeometryRef());
}
```

11.2.3.18　GetGeometriesInView

获取指定外包框矩形内的几何对象。在方法内部也是先设置图层的 SetSpatialFilterRect 属性完成要素的空间过滤，最后对每个符合过滤器要求的要素，调用 ParseOgrGeometry 方法将其转化为 SharpMap 的 Geometry 对象，代码如下：

```csharp
public Collection<Geometry> GetGeometriesInView(BoundingBox bbox)
{
    Collection<Geometry> geoms = new Collection<Geometry>();
    _ogrLayer.SetSpatialFilterRect(bbox.Left, bbox.Bottom, bbox.Right, bbox.Top);
    _ogrLayer.ResetReading();
    OgrFeature ogrFeature = null;
    try
    {
        while((ogrFeature = _ogrLayer.GetNextFeature())! = null)
        {
            Geometry geom = ParseOgrGeometry(ogrFeature.GetGeometryRef());
            if(geom ! = null) geoms.Add(geom);
            ogrFeature.Dispose();
        }
    }
    catch(Exception ex)
    {
        Debug.WriteLine(ex.Message);
    }
```

```
            return geoms;
    }
```

11.2.3.19　ExecuteIntersectionQuery

指向相交查询，将指定外包框矩形内的要素加入到参数中。这里也调用图层的 SetSpatialFilterRect 方法，将设置过滤器并调用 ExecuteIntersectionQuery 另一重载版本执行查询，代码如下：

```
public void ExecuteIntersectionQuery(BoundingBox bbox, FeatureDataSet ds)
{
    _ogrLayer.SetSpatialFilterRect(bbox.Left, bbox.Bottom, bbox.Right,
    bbox.Top);
    ExecuteIntersectionQuery(ds);
}
```

11.2.3.20　ExecuteIntersectionQuery

这里通过返回指定几何对象内的要素，这里将 SharpMap 类型的几何要素转化为 Ogr 类型的要素，作为空间过滤器：先调用 GeometryToWKB.Write（geom）方法将 SharpMap 类型的几何对象变成二进制几何对象，然后调用 OgrGeometry.CreateFromWkb（）方法将这个二进制对象转化为 Ogr 类型的几何对象，然后调用图层的 SetSpatialFilter 方法，将这个转换后的几何对象设置为空间过滤器，最后使用 ExecuteIntersectionQuery 方法的另一重载版本执行查询，代码如下：

```
public void ExecuteIntersectionQuery(Geometry geom, FeatureDataSet ds)
{
    OgrGeometry ogrGeometry =
    OgrGeometry.CreateFromWkb(GeometryToWKB.Write(geom));
    _ogrLayer.SetSpatialFilter(ogrGeometry);
    ExecuteIntersectionQuery(ds);
}
```

11.2.3.21　ExecuteIntersectionQuery

执行相交查询，上面两个重载方法都是调用此方法执行查询。这里先判断自定义的查询字符串是否为空，为空则调用图层的 SetAttributeFilter 方法设置为空字符串_ogrLayer.SetAttributeFilter（""）；若不为空则设置为自定义的查询字符串。然后调用图层的 ResetReading 方法重置读操作，构建一个 FeatureDataTable 类型的对象 myDt 用于保存查询结果，这里先调用 ReadColumnDefinition 方法读取图层的属性信息，参数为 myDt 和当前图层_ogrLayer，然后逐一读取查询结果，调用 OgrFeatureToFeatureDataRow 方法将 Ogr 类型的要素转化为 SharpMap 类型要素数据行，并加入要素数据表 myDt 中。最后将此要素数据表加入要素数据集中，代码如下：

```
private void ExecuteIntersectionQuery(FeatureDataSet ds)
{
    if (! String.IsNullOrEmpty(_definitionQuery))
```

11.2 OgrProvider 类

```
        _ogrLayer.SetAttributeFilter(_definitionQuery);
    else
        _ogrLayer.SetAttributeFilter("");
    _ogrLayer.ResetReading();
    //reads the column definition of the layer/feature
    FeatureDataTable myDt = new FeatureDataTable();
    ReadColumnDefinition(myDt, _ogrLayer);
    OgrFeature ogrFeature;
    while ((ogrFeature = _ogrLayer.GetNextFeature()) != null)
    {
        FeatureDataRow fdr = OgrFeatureToFeatureDataRow(myDt, ogrFeature);
        myDt.AddRow(fdr);
    }
    ds.Tables.Add(myDt);
}
```

11.2.3.22 ReadColumnDefinition

读取图层的属性信息。通过调用图层的 GetLayerDefn 方法可以得到图层属性相关信息，然后遍历每个属性，对于不同类型的属性，这里都将其转换为 SharpMap 对应类型的属性，而属性的名称则保持不变。如：对于 OgrFieldType.OFTInteger 类型的属性，这里通过反射生成一个 System.Int32 类型的属性，名称则通过 ogrFldDef.GetName() 方法获取，保持不变，代码如下：

```
private static void ReadColumnDefinition(FeatureDataTable fdt, OgrLayer oLayer)
{
    using (OgrFeatureDefn ogrFeatureDefn = oLayer.GetLayerDefn())
    {
        int iField;
        for (iField = 0; iField < ogrFeatureDefn.GetFieldCount(); iField++)
        {
            using (OgrFieldDefn ogrFldDef = ogrFeatureDefn.GetFieldDefn(iField))
            {
                OgrFieldType type = ogrFldDef.GetFieldType();
                switch (type)
                {
                    case OgrFieldType.OFTInteger:
                        fdt.Columns.Add(ogrFldDef.GetName(),
                            Type.GetType("System.Int32"));
                        break;
                    case OgrFieldType.OFTReal:
                        fdt.Columns.Add(ogrFldDef.GetName(),
                            Type.GetType("System.Double"));
                        break;
```

```csharp
                    case OgrFieldType.OFTString:
                        fdt.Columns.Add(ogrFldDef.GetName(),
                            Type.GetType("System.String"));
                        break;
                    case OgrFieldType.OFTWideString:
                        fdt.Columns.Add(ogrFldDef.GetName(),
                            Type.GetType("System.String"));
                        break;
                    case OgrFieldType.OFTDate:
                    case OgrFieldType.OFTTime:
                    case OgrFieldType.OFTDateTime:
                        fdt.Columns.Add(ogrFldDef.GetName(), typeof(DateTime));
                        break;
                    default:
                        {
                            //fdt.Columns.Add(_OgrFldDef.GetName(),
                            System.Type.GetType("System.String"));
                            Debug.WriteLine("Not supported type:" + type + "["
                                + ogrFldDef.GetName() + "]");
                            break;
                        }
                }
            }
        }
    }
}
```

11.2.3.23 ParseOgrGeometry

将 Ogr 类型的几何对象转换为 SharpMap 类型的几何对象，转换之前调用几何对象的 FlattenTo2D 方法，将 wkbPoint25D 类型（含有 z 坐标的点）转换为 2D 的类型。每个 2D 的几何图元类型都对应一个 2.5D 的编码，因此我们在代码中可能需要对 2D 或 3D 情形做适当的处理。

构建一个几何对象 WkbSize 方法返回值大小的 byte [] 数组对象 wkbBuffer，然后将调用几何对象的 ExportToWkb 方法将集合对象导入到 byte [] 数组中，用 GeometryFromWKB 类的 parse 方法将此数组转换为 SharpMap 类型的集合对象并返回，代码如下：

```csharp
private static Geometry ParseOgrGeometry(OgrGeometry ogrGeometry)
{
    if (ogrGeometry != null)
    {
        //Just in case it isn't 2D
        ogrGeometry.FlattenTo2D();
        byte[] wkbBuffer = new byte[ogrGeometry.WkbSize()];
```

11.2 OgrProvider 类

```
            ogrGeometry.ExportToWkb(wkbBuffer);
            Geometry geom = GeometryFromWKB.Parse(wkbBuffer);
            if(geom = = null)
                Debug.WriteLine(string.Format("Failed to parse '{0}'",
                    ogrGeometry.GetGeometryType()));
            return geom;
        }
        return null;
    }
```

11.2.3.24 OgrFeatureToFeatureDataRow

将 Ogr 类型的几何对象加入到 SharpMap 类型的要素数据表中,这里先生成一个新的要素数据行 fdr,用于存放几何对象,对于不同类型的属性调用几何对象的 GetFieldAsString、GetFieldAsInteger、GetFieldAsDouble 或者 GetFieldAsDateTime 方法获取对应类型的值。可以看出,这里不支持 OgrFieldType.OFTStringList、OgrFieldType.OFTWideStringList、OgrFieldType.OFTIntegerList 类型的数据,代码如下:

```
    private static FeatureDataRow OgrFeatureToFeatureDataRow(FeatureDataTable table, OS-
Geo.OGR.Feature ogrFeature)
    {
        FeatureDataRow fdr = table.NewRow();
        Int32 fdrIndex = 0;
        for(int iField = 0; iField < ogrFeature.GetFieldCount(); iField + +)
        {
            switch(ogrFeature.GetFieldType(iField))
            {
                case OgrFieldType.OFTString:
                case OgrFieldType.OFTWideString:
                    fdr[fdrIndex + +] = ogrFeature.GetFieldAsString(iField);
                    break;
                case OgrFieldType.OFTStringList:
                case OgrFieldType.OFTWideStringList:
                    break;
                case OgrFieldType.OFTInteger:
                    fdr[fdrIndex + +] = ogrFeature.GetFieldAsInteger(iField);
                    break;
                case OgrFieldType.OFTIntegerList:
                    break;
                case OgrFieldType.OFTReal:
                    fdr[fdrIndex + +] = ogrFeature.GetFieldAsDouble(iField);
                    break;
                case OgrFieldType.OFTRealList:
                    break;
```

```
            case OgrFieldType.OFTDate:
            case OgrFieldType.OFTDateTime:
            case OgrFieldType.OFTTime:
                Int32 y, m, d, h, mi, s, tz;
                ogrFeature.GetFieldAsDateTime(iField, out y, out m, out d, out h,
                    out mi, out s, out tz);
                fdr[fdrIndex++] = new DateTime(y, m, d, h, mi, s);
                break;
            default:
                Debug.WriteLine(string.Format("Cannot handle Ogr DataType '{0}'",
                    ogrFeature.GetFieldType(iField)));
                break;
        }
    }
    fdr.Geometry = ParseOgrGeometry(ogrFeature.GetGeometryRef());
    return fdr;
}
```

11.2.3.25 ExecuteQuery(string query)

执行查询。这里设置几何对象查询过滤器参数为空,调用另一重载版本,代码如下:

```
public FeatureDataSet ExecuteQuery(string query)
{
    return ExecuteQuery(query, null);
}
```

11.2.3.26 ExecuteQuery(string query, OgrGeometry filter)

执行查询。这里将查询字符串作为参数,调用数据源的 ExecuteSQL 方法执行查询,并用相交查询类似的方法输出查询结果,代码如下:

```
public FeatureDataSet ExecuteQuery(string query, OgrGeometry filter)
{
    try
    {
        FeatureDataSet ds = new FeatureDataSet();
        FeatureDataTable myDt = new FeatureDataTable();
        OgrLayer results = _ogrDataSource.ExecuteSQL(query, filter, "");
        //reads the column definition of the layer/feature
        ReadColumnDefinition(myDt, results);
        OgrFeature ogrFeature;
        results.ResetReading();
        while ((ogrFeature = results.GetNextFeature()) != null)
        {
            FeatureDataRow dr = OgrFeatureToFeatureDataRow(myDt, ogrFeature);
```

```
            myDt. AddRow(dr);
        }
        ds. Tables. Add(myDt);
        _ogrDataSource. ReleaseResultSet(results);
        return ds;
    }
    catch (Exception exc)
    {
        Debug. WriteLine(exc. ToString());
        return new FeatureDataSet();
    }
}
```

11.2.3.27　QueryFeatures

执行查询，返回指定几何对象周围指定距离（缓冲区内）内的集合对象。这里并未实现，抛出未实现异常，代码如下：

```
[Obsolete("Use ExecuteIntersectionQuery instead")]
public FeatureDataTable QueryFeatures(Geometry geom, double distance)
{
    throw new NotSupportedException();
}
```

11.2.3.28　ExecuteIntersectionQuery

执行相交查询，参数为指定的集合对象。这里直接调用另一重载版本，代码如下：

```
public FeatureDataTable ExecuteIntersectionQuery(Geometry geom)
{
    FeatureDataSet fds = new FeatureDataSet();
    ExecuteIntersectionQuery(geom, fds);
    return fds. Tables[0];
}
```

11.2.3.29　GetFeaturesInView

得到指定外包框中的几何对象。这里也是调用另外的重载版本，详细信息请参照上述解释，代码如下：

```
[Obsolete("Use ExecuteIntersectionQuery(BoundingBox,FeatureDataSet) instead")]
public void GetFeaturesInView(BoundingBox bbox, FeatureDataSet ds)
{
    ExecuteIntersectionQuery(bbox, ds);
}
```

11.3　GdalRasterLayer 类

在 SharMap 的基本类库中，只有矢量图层 VectorLayer，而没有用于表达栅格数据的图层。在 SharpMap 的扩展库中，定义了可以访问 GDAL 栅格数据的图层类。由于该类的实现代码与栅格数据格式紧密相关，本章只对代码做简单陈述，有些专业术语的含义请读者结合相关栅格数据格式资料自行研究。

在示例程序 WinFormSamples 项目中使用了 GdalRasterLayer 类，GdalSample、OgrSample 类中使用该类表达 GDAL/OGR 数据。

11.3.1 GdalRasterLayer 类的数据成员

11.3.1.1 _gdalDataset
表示 gdal 数据集，代码如下：

```
protected Dataset _gdalDataset;
```

11.3.1.2 _geoTransform
表示地理变换，代码如下：

```
internal GeoTransform _geoTransform;
```

11.3.1.3 disposed
表示资源是否已经被回收，代码如下：

```
private bool disposed;
```

11.3.2 GdalRasterLayer 类的属性

11.3.2.1 FWToolsVersion
获取用 FWTools 程序的版本信息，代码如下：

```
public static string FWToolsVersion
{
    get { return FWToolsHelper.FWToolsVersion; }
}
```

11.3.2.2 Filename
设置和获取栅格数据的文件名，代码如下：

```
private string _Filename;
public string Filename
{
    get { return _Filename; }
    set { _Filename = value; }
}
```

11.3.2.3 BitDepth
获取和设置位深。栅格数据的位深，又称像素深度或位深度，用于确定存储在各波段中的可能值的范围。八位深度将存储 2^8 = 256 个值（0 到 255），而 16 位深度则会存储

$2^{16} = 65536$ 个值（0 到 65535），这里设置默认值为 8，代码如下：

```csharp
private int _bitDepth = 8;
public int BitDepth
{
    get { return _bitDepth; }
    set { _bitDepth = value; }
}
```

11.3.2.4　Projection

设置和获取投影，代码如下：

```csharp
private string _projectionWkt = "";
public string Projection
{
    get { return _projectionWkt; }
    set { _projectionWkt = value; }
}
```

11.3.2.5　DisplayIR

获取和设置是否显示红外波段，代码如下：

```csharp
private bool _displayIR;
public bool DisplayIR
{
    get { return _displayIR; }
    set { _displayIR = value; }
}
```

11.3.2.6　DisplayCIR

获取和设置是否显示彩色红外，代码如下：

```csharp
private bool _displayCIR;
public bool DisplayCIR
{
    get { return _displayCIR; }
    set { _displayCIR = value; }
}
```

11.3.2.7　ShowClip

获取或设置是否显示裁减，代码如下：

```csharp
private bool _showClip;
public bool ShowClip
{
    get { return _showClip; }
```

```csharp
    set { _showClip = value; }
}
```

11.3.2.8 Gamma

获取和设置伽玛值，代码如下：

```csharp
private double _gamma = 1;
public double Gamma
{
    get { return _gamma; }
    set { _gamma = value; }
}
```

11.3.2.9 SpotGamma

获取和设置_spotGamma，代码如下：

```csharp
private double _spotGamma = 1;
public double SpotGamma
{
    get { return _spotGamma; }
    set { _spotGamma = value; }
}
```

11.3.2.10 NonSpotGamma

获取和设置_nonSpotGamma，代码如下：

```csharp
private double _nonSpotGamma = 1;
public double NonSpotGamma
{
    get { return _nonSpotGamma; }
    set { _nonSpotGamma = value; }
}
```

11.3.2.11 Gain

获取和设置增益数组，代码如下：

```csharp
private double[] _gain = { 1, 1, 1, 1 };
public double[] Gain
{
    get { return _gain; }
    set { _gain = value; }
}
```

11.3.2.12 SpotGain

获取和设置_spotGain，代码如下：

```csharp
private double[] _spotGain = { 1, 1, 1, 1 };
```

11.3 GdalRasterLayer 类

```csharp
public double[] SpotGain
{
    get { return _spotGain; }
    set { _spotGain = value; }
}
```

11.3.2.13 NonSpotGain

获取和设置_nonSpotGain，代码如下：

```csharp
private double[] _nonSpotGain = { 1, 1, 1, 1 };
public double[] NonSpotGain
{
    get { return _nonSpotGain; }
    set { _nonSpotGain = value; }
}
```

11.3.2.14 CurveLut

获取和设置曲线纠正，代码如下：

```csharp
private List<int[]> _curveLut;
public List<int[]> CurveLut
{
    get { return _curveLut; }
    set { _curveLut = value; }
}
```

11.3.2.15 HaveSpot

获取和设置是否有污点，代码如下：

```csharp
private bool _haveSpot;
public bool HaveSpot
{
    get { return _haveSpot; }
    set { _haveSpot = value; }
}
```

11.3.2.16 SpotCurveLut

获取和设置_spotCurveLut，代码如下：

```csharp
private List<int[]> _spotCurveLut;
public List<int[]> SpotCurveLut
{
    get { return _spotCurveLut; }
    set { _spotCurveLut = value; }
}
```

11.3.2.17 NonSpotCurveLut

获取和设置_nonSpotCurveLut,代码如下:

```csharp
private List<int[]> _nonSpotCurveLut;
public List<int[]> NonSpotCurveLut
{
    get { return _nonSpotCurveLut; }
    set { _nonSpotCurveLut = value; }
}
```

11.3.2.18 SpotPoint

获取和设置点对象,代码如下:

```csharp
private PointF _spot = new PointF(0, 0);
public PointF SpotPoint
{
    get { return _spot; }
    set { _spot = value; }
}
```

11.3.2.19 InnerSpotRadius

获取和设置内点半径,代码如下:

```csharp
private double _innerSpotRadius;
public double InnerSpotRadius
{
    get { return _innerSpotRadius; }
    set { _innerSpotRadius = value; }
}
```

11.3.2.20 OuterSpotRadius

获取和设置_outerSpotRadius,代码如下:

```csharp
private double _outerSpotRadius;
public double OuterSpotRadius
{
    get { return _outerSpotRadius; }
    set { _outerSpotRadius = value; }
}
```

11.3.2.21 Histogram

获取和设置统计直方图,代码如下:

```csharp
private List<int[]> _histogram;
public List<int[]> Histogram
{
    get { return _histogram; }
}
```

11.3.2.22 HistoMean

获取和设置平均值,代码如下:

```
private double[ ] _histoMean;
public double[ ] HistoMean
{
    get { return _histoMean; }
}
```

11.3.2.23 HistoBrightness

获取和设置统计直方图亮度值,代码如下:

```
private double _histoBrightness;
public double HistoBrightness
{
    get { return _histoBrightness; }
}
```

11.3.2.24 HistoContrast

获取和设置统计直方图对比度,代码如下:

```
private double _histoContrast;
public double HistoContrast
{
    get { return _histoContrast; }
}
```

11.3.2.25 Bands

获取和设置波段数,代码如下:

```
internal int _lbands;
public int Bands
{
    get { return _lbands; }
}
```

11.3.2.26 GSD

获取和设置 GSD 值(水平像素分辨率)。代码如下:

```
public double GSD
{
    get { return _geoTransform.HorizontalPixelResolution; }
}
```

11.3.2.27 UseRotation

获取和设置是否使用旋转,代码如下:

```csharp
            protected bool _useRotation = true;
            public bool UseRotation
            {
                get { return _useRotation; }
                set
                {
                    _useRotation = value;
                    _envelope = GetExtent();
                }
            }
```

11.3.2.28 Size

获取和设置图像大小,代码如下:

```csharp
            protected Size _imagesize;
            public Size Size
            {
                get { return _imagesize; }
            }
```

11.3.2.29 ColorCorrect

获取和设置颜色纠正,代码如下:

```csharp
            private bool _colorCorrect = true;
            public bool ColorCorrect
            {
                get { return _colorCorrect; }
                set { _colorCorrect = value; }
            }
```

11.3.2.30 HistoBounds

获取和设置统计直方图边界,代码如下:

```csharp
            private Rectangle _histoBounds;
            public Rectangle HistoBounds
            {
                get { return _histoBounds; }
                set { _histoBounds = value; }
            }
```

11.3.2.31 Transform

获取和设置坐标变换,代码如下:

```csharp
            protected ICoordinateTransformation _transform;
            public ICoordinateTransformation Transform
            {
```

11.3.2.32 TransparentColor

获取和设置透明色,代码如下:

```
private Color _transparentColor
public Color TransparentColor
{
    get { return _transparentColor; }
    set { _transparentColor = value; }
}
```

11.3.2.33 StretchPoint

获取和设置延伸点,代码如下:

```
private Point _stretchPoint;
public Point StretchPoint
{
    get
    {
        if (_stretchPoint.Y == 0)
            ComputeStretch();
        return _stretchPoint;
    }
    set { _stretchPoint = value; }
}
```

11.3.2.34 Envelope

获取和设置外包框矩形,代码如下:

```
protected BoundingBox _envelope;
public override BoundingBox Envelope
{
    get { return _envelope; }
}
```

11.3.2.35 IsQueryEnabled

获取和设置图层是否能够被查询的值,代码如下:

```
private bool _isQueryEnabled = true;
public bool IsQueryEnabled
{
    get { return _isQueryEnabled; }
    set { _isQueryEnabled = value; }
}
```

11.3.2.36 NoDataInitColor

获取和设置无数据初始化颜色，代码如下：

```csharp
private Color _noDataInitColor = Color.Yellow;
public Color NoDataInitColor
{
    get { return _noDataInitColor; }
    set { _noDataInitColor = value; }
}
```

11.3.2.37 ColorBlend

获取和设置颜色混合，代码如下：

```csharp
private ColorBlend _colorBlend;
public ColorBlend ColorBlend
{
    get { return _colorBlend; }
    set { _colorBlend = value; }
}
```

11.3.3 GdalRasterLayer 类的方法

11.3.3.1 GdalRasterLayer（）

静态构造函数。这里调用 FwToolsHelper.Configure（）方法设置配置信息，代码如下：

```csharp
static GdalRasterLayer()
{
    FwToolsHelper.Configure();
}
```

11.3.3.2 GdalRasterLayer（string strLayerName，string imageFilename）

构造函数，需传入图层的名称、图像的文件名然后调用 Gdal.AllRegister（）方法注册驱动，然后调用 GDAL.OpenShared 方法打开栅格数据集，保存在成员变量 _gdalDataset 中，代码如下：

```csharp
_gdalDataset = Gdal.OpenShared(_Filename, Access.GA_ReadOnly);
```

然后调用数据集的 GetProjectionRef 获取投影信息：

```csharp
_projectionWkt = _gdalDataset.GetProjectionRef();
```

以上返回表示投影信息的字符串。若字符串为空，表示数据集中没有相应的投影，就需要查找数据文件路径下对应名称相同且以 ".prg" 为后缀的文件（投影文件），若存在这样的文件，则调用 File.ReadAllText 方法将投影文件的信息读取到字符串中，并赋予变量 _projectionWkt：

11.3 GdalRasterLayer 类

```
_projectionWkt = File.ReadAllText(imageFilename.Substring(0, imageFilename.LastIndexOf
(".")) + ".prj");
```

然后设置图像的尺寸为数据集对应的尺寸，使用的是数据集的 RasterXSize 属性和 RasterYSize 属性，分别代表了数据集在 X 和 Y 方向上的大小：

```
_imagesize = new Size(_gdalDataset.RasterXSize, _gdalDataset.RasterYSize);
```

接下来调用 GetExtent() 方法设置外包框矩形并赋予变量 _envelope，GetExtent 方法后面会有解释：

```
_envelope = GetExtent();
```

设置统计直方图的统计范围为外包框矩形对应的大小：

```
_histoBounds = new Rectangle((int)_envelope.Left, (int)_envelope.Bottom, (int)_envelope.Width, (int)_envelope.Height);
```

最后，通过数据集的 RasterCount 属性获取图像的波段数，赋予变量 _lbands：

```
_lbands = _gdalDataset.RasterCount;
```

完整代码如下：

```
public GdalRasterLayer(string strLayerName, string imageFilename)
{
    LayerName = strLayerName;
    Filename = imageFilename;
    disposed = false;
    Gdal.AllRegister();
    try
    {
        _gdalDataset = Gdal.OpenShared(_Filename, Access.GA_ReadOnly);
        _projectionWkt = _gdalDataset.GetProjectionRef();
        if (_projectionWkt == "" &&
            File.Exists(imageFilename.Substring(0,
            imageFilename.LastIndexOf(".")) + ".prj"))
        {
            _projectionWkt =
                File.ReadAllText(imageFilename.Substring(0,
                imageFilename.LastIndexOf(".")) + ".prj");
        }
        _imagesize = new Size(_gdalDataset.RasterXSize,
        _gdalDataset.RasterYSize);
        _envelope = GetExtent();
        _histoBounds = new Rectangle((int)_envelope.Left, (int)_envelope.Bottom,
        (int)_envelope.Width, (int)_envelope.Height);
        _lbands = _gdalDataset.RasterCount;
```

```
        catch(Exception ex)
        {
            _gdalDataset = null;
            throw new Exception("Couldn't load " + imageFilename + "\n\n" + ex.Message
                + ex.InnerException);
        }
    }
```

11.3.3.3 Dispose (bool disposing)

资源清理方法,如果资源没有释放,就需调用_gdalDataset.Dispose()方法完成资源的清理工作,代码如下:

```
private void Dispose(bool disposing)
{
    if(! disposed)
    {
        if(disposing)
            if(_gdalDataset != null)
            {
                try
                {_gdalDataset.Dispose();}
                finally{_gdalDataset = null;}
            }
        disposed = true;
    }
}
```

11.3.3.4 Dispose

资源清理方法。这里调用.NET 的垃圾回收器进行一次资源回收工作,代码如下:

```
public void Dispose()
{
    Dispose(true);
    GC.SuppressFinalize(this);
}
```

11.3.3.5 ~GdalRasterLayer()

析构函数。这里主要调用 dispose()方法,代码如下:

```
~GdalRasterLayer()
{
    Dispose(true);
}
```

11.3.3.6 ApplyColorCorrection

应用色彩纠正,传入的参数为影像的值 imageVal、污点值 spotVal、色彩通道 channel、点的投影坐标 X、点的投影坐标 Y。首先判断是否有污点,若有污点则根据_nonSpotGamma 的值是否为 1,否则重新设置像元值,代码如下:

```
imageVal = 256 * Math.Pow((imageVal / 256), _nonSpotGamma);
```

根据色彩通道进行不同的增益变换:

```
if (channel == 2) imageVal = imageVal * _nonSpotGain[0];
else if (channel == 1) imageVal = imageVal * _nonSpotGain[1];
else if (channel == 0) imageVal = imageVal * _nonSpotGain[2];
else if (channel == 3) imageVal = imageVal * _nonSpotGain[3];
```

最后,若像元的值超过 255,则设置为 255:

```
if (imageVal > 255) imageVal = 255;
```

完整代码如下:

```
private double ApplyColorCorrection(double imageVal, double spotVal, int channel, double GndX, double GndY)
{
    double finalVal;
    double distance;
    double imagePct, spotPct;
    finalVal = imageVal;
    if (_haveSpot)
    {
        // gamma
        if (_nonSpotGamma != 1)
            imageVal = 256 * Math.Pow((imageVal / 256), _nonSpotGamma);
        // gain
        if (channel == 2)
            imageVal = imageVal * _nonSpotGain[0];
        else if (channel == 1)
            imageVal = imageVal * _nonSpotGain[1];
        else if (channel == 0)
            imageVal = imageVal * _nonSpotGain[2];
        else if (channel == 3)
            imageVal = imageVal * _nonSpotGain[3];
        if (imageVal > 255)
            imageVal = 255;
        // curve
        if (_nonSpotCurveLut != null)
            if (_nonSpotCurveLut.Count != 0)
```

```
            if (channel == 2 || channel == 4)
                imageVal = _nonSpotCurveLut[0][(int)imageVal];
            else if (channel == 1)
                imageVal = _nonSpotCurveLut[1][(int)imageVal];
            else if (channel == 0)
                imageVal = _nonSpotCurveLut[2][(int)imageVal];
            else if (channel == 3)
                imageVal = _nonSpotCurveLut[3][(int)imageVal];
    }
finalVal = imageVal;
distance = Math.Sqrt(Math.Pow(GndX - SpotPoint.X, 2) + Math.Pow(GndY - SpotPoint.Y, 2));
if (distance <= _innerSpotRadius + _outerSpotRadius)
{
    // gamma
    if (_spotGamma != 1)
        spotVal = 256 * Math.Pow((spotVal / 256), _spotGamma);
    // gain
    if (channel == 2)
        spotVal = spotVal * _spotGain[0];
    else if (channel == 1)
        spotVal = spotVal * _spotGain[1];
    else if (channel == 0)
        spotVal = spotVal * _spotGain[2];
    else if (channel == 3)
        spotVal = spotVal * _spotGain[3];
    if (spotVal > 255)
        spotVal = 255;
    // curve
    if (_spotCurveLut != null)
        if (_spotCurveLut.Count != 0)
        {
            if (channel == 2 || channel == 4)
                spotVal = _spotCurveLut[0][(int)spotVal];
            else if (channel == 1)
                spotVal = _spotCurveLut[1][(int)spotVal];
            else if (channel == 0)
                spotVal = _spotCurveLut[2][(int)spotVal];
            else if (channel == 3)
                spotVal = _spotCurveLut[3][(int)spotVal];
        }
    if (distance < _innerSpotRadius)
```

```
                finalVal = spotVal;
            else
            {
                imagePct = (distance - _innerSpotRadius) / _outerSpotRadius;
                spotPct = 1 - imagePct;
                finalVal = (Math.Round((spotVal * spotPct) + (imageVal *
                    imagePct)));
            }
        }
    }
    // gamma
    if (_gamma != 1)
        finalVal = (256 * Math.Pow((finalVal / 256), _gamma));
    switch (channel)
    {
        case 2:
            finalVal = finalVal * _gain[0];
            break;
        case 1:
            finalVal = finalVal * _gain[1];
            break;
        case 0:
            finalVal = finalVal * _gain[2];
            break;
        case 3:
            finalVal = finalVal * _gain[3];
            break;
    }
    if (finalVal > 255)
        finalVal = 255;
    // curve
    if (_curveLut != null)
        if (_curveLut.Count != 0)
        {
            if (channel == 2 || channel == 4)
                finalVal = _curveLut[0][(int)finalVal];
            else if (channel == 1)
                finalVal = _curveLut[1][(int)finalVal];
            else if (channel == 0)
                finalVal = _curveLut[2][(int)finalVal];
            else if (channel == 3)
                finalVal = _curveLut[3][(int)finalVal];
        }
```

```
            return finalVal;
    }
```

11.3.3.7 ApplyTransformToEnvelope

应用投影变换，需要变换数据集的外包矩形框和统计直方图范围框。这里首先声明四个数组变量，用于保存统计直方图统计范围的四个角点坐标，声明四个 double 类型的变量保存变换后的边界值，代码如下：

```
double[] leftBottom, leftTop, rightTop, rightBottom;
double left, right, bottom, top;
```

调用 GetExtent 方法得到数据集的外包框矩形，保存于成员变量_envelope 中：

```
_envelope = GetExtent();
```

将变换后的外包框赋予成员变量_envelope：

```
_envelope = GeometryTransform.TransformBox(_envelope, _transform.MathTransform);
```

然后获取统计直方图的边界值，将结果保存于事先声明的变量 leftBottom、leftTop、rightTop、rightBottom 中：

```
// do same to histo rectangle
leftBottom = new double[] { _histoBounds.Left, _histoBounds.Bottom };
leftTop = new double[] { _histoBounds.Left, _histoBounds.Top };
rightBottom = new double[] { _histoBounds.Right, _histoBounds.Bottom };
rightTop = new double[] { _histoBounds.Right, _histoBounds.Top };
```

对边界点做投影变换，并将结果保存：

```
// transform corners into new projection
leftBottom = _transform.MathTransform.Transform(leftBottom);
leftTop = _transform.MathTransform.Transform(leftTop);
rightBottom = _transform.MathTransform.Transform(rightBottom);
rightTop = _transform.MathTransform.Transform(rightTop);
```

然后计算边界值，保存到变量 left、right、bottom、top 中：

```
// find extents
left = Math.Min(leftBottom[0], Math.Min(leftTop[0], Math.Min(rightBottom[0], rightTop[0])));
right = Math.Max(leftBottom[0], Math.Max(leftTop[0], Math.Max(rightBottom[0], rightTop[0])));
bottom = Math.Min(leftBottom[1], Math.Min(leftTop[1], Math.Min(rightBottom[1], rightTop[1])));
top = Math.Max(leftBottom[1], Math.Max(leftTop[1], Math.Max(rightBottom[1], rightTop[1])));
```

最后用边界值生成新的外包框矩形，保存到成员变量_histoBounds 中：

11.3 GdalRasterLayer 类

```
// set histo rectangle
_histoBounds = new Rectangle((int)left, (int)bottom, (int)right, (int)top);
```

完整代码如下：

```
// applies map projection transfrom to get reprojected envelope
private void ApplyTransformToEnvelope()
{
    double[] leftBottom, leftTop, rightTop, rightBottom;
    double left, right, bottom, top;
    _envelope = GetExtent();
    if (_transform == null)
        return;
    // set envelope
    _envelope = GeometryTransform.TransformBox(_envelope,
    _transform.MathTransform);
    // do same to histo rectangle
    leftBottom = new double[] { _histoBounds.Left, _histoBounds.Bottom };
    leftTop = new double[] { _histoBounds.Left, _histoBounds.Top };
    rightBottom = new double[] { _histoBounds.Right, _histoBounds.Bottom };
    rightTop = new double[] { _histoBounds.Right, _histoBounds.Top };
    // transform corners into new projection
    leftBottom = _transform.MathTransform.Transform(leftBottom);
    leftTop = _transform.MathTransform.Transform(leftTop);
    rightBottom = _transform.MathTransform.Transform(rightBottom);
    rightTop = _transform.MathTransform.Transform(rightTop);
    // find extents
    left = Math.Min(leftBottom[0], Math.Min(leftTop[0], Math.Min(rightBottom[0],
    rightTop[0])));
    right = Math.Max(leftBottom[0], Math.Max(leftTop[0], Math.Max(rightBottom[0],
    rightTop[0])));
    bottom = Math.Min(leftBottom[1], ath.Min(leftTop[1], Math.Min(rightBottom[1],
    rightTop[1])));
    top = Math.Max(leftBottom[1], Math.Max(leftTop[1], Math.Max(rightBottom[1],
    rightTop[1])));
    // set histo rectangle
    _histoBounds = new Rectangle((int)left, (int)bottom, (int)right, (int)top);
}
```

11.3.3.8 ComputeStretch

计算拉伸，当图像需要拉伸时，运用此函数可以计算拉伸点（X 值表示最大值，Y 值表示最小值），先判断数据集的宽度是否大于 4000，若不大于则完全读取图像，若大于则只读取宽度在 4000 以内的图像（长度为长宽比保持不变时对应的值），将要读取的长度、宽度保存到变量 width、height 中，代码如下：

```
    if ( _gdalDataset. RasterYSize < 4000)
    {
        height = _gdalDataset. RasterYSize;
        width = _gdalDataset. RasterXSize;
    }
    else
    {
        height = 4000;
        width = (int)(4000 * (_gdalDataset. RasterXSize /
            (double)_gdalDataset. RasterYSize));
    }
```

对于数据集中的每个波段,找到在将要读取的长度、宽度内的最大、最小值,并对不同位深的图像计算出对应的十进制值,最后以最小值为 X、最大值为 Y 生成点对象,保存到成员变量_stretchPoint 中。完整代码如下:

```
    // find min and max pixel values of the image
    private void ComputeStretch( )
    {
        double min = 99999999, max = -99999999;
        int width, height;
        if ( _gdalDataset. RasterYSize < 4000)
        {
            height = _gdalDataset. RasterYSize;
            width = _gdalDataset. RasterXSize;
        }
        else
        {
            height = 4000;
            width = (int)(4000 * (_gdalDataset. RasterXSize /
                (double)_gdalDataset. RasterYSize));
        }
        double[ ] buffer = new double[ width * height];
        for (int band = 1; band < = _lbands; band + +)
        {
            Band RBand = _gdalDataset. GetRasterBand(band);
            RBand. ReadRaster(0, 0, _gdalDataset. RasterXSize,
                _gdalDataset. RasterYSize, buffer, width, height, 0, 0);
            for (int i = 0; i < buffer. Length; i + +)
            {
                if (buffer[i] < min) min = buffer[i];
                if (buffer[i] > max) max = buffer[i];
            }
        }
```

```
if ( _bitDepth = = 12) {min / = 16; max / = 16;}
else if ( _bitDepth = = 16) {min / = 256; max / = 256;}
if ( max > 255) max = 255;
_stretchPoint = new Point((int)min, (int)max);
}
```

11.3.3.9 ExecuteIntersectionQuery

执行相交查询，返回指定外包框内的要素数据集，这里构建外包框矩形的中心点，代码如下：

```
Geometries.Point pt = new Geometries.Point( box.Left + 0.5 * box.Width, box.Top - 0.5 * box.Height);
```

调用另一重载版本执行查询：

```
public void ExecuteIntersectionQuery( BoundingBox box, FeatureDataSet ds)
{
    Geometries.Point pt = new Geometries.Point( box.Left + 0.5 * box.Width,
                                                box.Top - 0.5 * box.Height);
    ExecuteIntersectionQuery( pt, ds);
}
```

11.3.3.10 ExecuteIntersectionQuery

执行相交查询，找到给定点对应所有波段的值并返回。这里若坐标变换不为空，先对点做坐标变换，代码如下：

```
if ( CoordinateTransformation ! = null)
{
    CoordinateTransformation.MathTransform.Invert();
    pt = GeometryTransform.TransformPoint( pt,
    CoordinateTransformation.MathTransform);
    CoordinateTransformation.MathTransform.Invert();
}
```

然后设置结果表的属性列，这里添加三个列 Ordinate X、Ordinate Y、Value Band {0}，都是 double 类型，分别表示 X 坐标、Y 坐标和对应波段的值：

```
FeatureDataTable dt = new FeatureDataTable();
dt.Columns.Add("Ordinate X", typeof(Double));
dt.Columns.Add("Ordinate Y", typeof(Double));
for ( int i = 1; i < = Bands; i + +)
    dt.Columns.Add( string.Format("Value Band {0}", i), typeof(Double));
```

将点变换到图像坐标，检查点是否在图像的范围内，若不在则直接返回：

```
Geometries.Point imgPt = _geoTransform.GroundToImage( pt);
Int32 x = Convert.ToInt32( imgPt.X);
```

```
Int32 y = Convert.ToInt32(imgPt.Y);
if (x < 0) return;
if (y < 0) return;
if (x >= _imagesize.Width) return;
if (y >= _imagesize.Height) return;
```

最后遍历每个图层,读取图层中对应点的值,将读取的结果保存到要素结果表中,对于没有值的点,保存时设置为 = Double.NaN;:

```
for (int i = 1; i <= Bands; i++)
{
    Band band = _gdalDataset.GetRasterBand(i);
    //DataType dtype = band.DataType;
    CPLErr res = band.ReadRaster(x, y, 1, 1, buffer, 1, 1, 0, 0);
    if (res == CPLErr.CE_None) { dr[1 + i] = buffer[0]; }
    else { dr[1 + i] = Double.NaN; }
}
dt.Rows.Add(dr);
ds.Tables.Add(dt);
```

11.3.3.11　GetExtent

获取图层的外包框矩形。若_gdalDataset 不为空,调用_gdalDataset 的 GetGeoTransform 方法获取仿射变换参数,保存于 double 类型的数组 geoTrans 中,若不需要旋转,且没有污点数据,或者 geoTrans [0]、geoTrans [3] 都为0,则新建一个数组,赋予 geoTrans;新建一个 GeoTransform 对象,用于地理变换,保存于_geoTransform 中;接着获取图像的高度和宽度,分别调用_geoTransform 对象的 EnvelopeLeft、EnvelopeRight、EnvelopeTop、EnvelopeBottom 方法进行转换,并将转换的结果作为参数,新建 BoundingBox 对象并返回,代码如下:

```
private BoundingBox GetExtent()
{
    if (_gdalDataset != null)
    {
        double right = 0, left = 0, top = 0, bottom = 0;
        double dblW, dblH;
        double[] geoTrans = new double[6];
        _gdalDataset.GetGeoTransform(geoTrans);
        // no rotation... use default transform
        if (!_useRotation && !_haveSpot || (geoTrans[0] == 0 && geoTrans[3] == 0))
            geoTrans = new[] { 999.5, 1, 0, 1000.5, 0, -1 };
        _geoTransform = new GeoTransform(geoTrans);
        // image pixels
        dblW = _imagesize.Width;
        dblH = _imagesize.Height;
```

11.3 GdalRasterLayer 类

```
            left = _geoTransform.EnvelopeLeft(dblW, dblH);
            right = _geoTransform.EnvelopeRight(dblW, dblH);
            top = _geoTransform.EnvelopeTop(dblW, dblH);
            bottom = _geoTransform.EnvelopeBottom(dblW, dblH);
            return new BoundingBox(left, bottom, right, top);
        }
        return null;
    }
```

11.3.3.12 GetFourCorners

获取图层的四个角点,这里首先新建一个点的集合对象 points。若_gdalDataset 不为空,先调用_gdalDataset 的 GetGeoTransform 方法获取仿射变换参数,保存于 double 类型的数组 geoTrans 中,若不需要旋转,且没有污点数据,或者 geoTrans[0]、geoTrans[3] 都为 0,则新建一个数组,赋予 geoTrans;然后将四个角点加入到点的集合 point 中,代码如下:

```
public Collection<Geometries.Point> GetFourCorners()
{
    Collection<Geometries.Point> points = new Collection<Geometries.Point>();
    double[] dblPoint;
    if (_gdalDataset != null)
    {
        double[] geoTrans = new double[6];
        _gdalDataset.GetGeoTransform(geoTrans);
        // no rotation... use default transform
        if (!_useRotation && !_haveSpot || (geoTrans[0] == 0 && geoTrans[3] == 0))
            geoTrans = new[] {999.5, 1, 0, 1000.5, 0, -1};
        points.Add(new Geometries.Point(geoTrans[0], geoTrans[3]));
        points.Add(new Geometries.Point(geoTrans[0] + (geoTrans[1] *
            _imagesize.Width), geoTrans[3] + (geoTrans[4] * _imagesize.Width)));
        points.Add(new Geometries.Point(geoTrans[0] + (geoTrans[1] * _imagesize.Width) +
            (geoTrans[2] * _imagesize.Height), geoTrans[3] + (geoTrans[4] *
            _imagesize.Width) + (geoTrans[5] * _imagesize.Height)));
        points.Add(new Geometries.Point(geoTrans[0] + (geoTrans[2] *
            _imagesize.Height), geoTrans[3] + (geoTrans[5] * _imagesize.Height)));
```

若_transform 不为空,遍历每个点,调用_transform.MathTransform.Transform 方法进行地理变换,并将变换后的点存入 point 集合中,变换完成后返回此 point 集合:

```
        if (_transform != null)
        {
            for (int i = 0; i < 4; i++)
            {
                dblPoint = _transform.MathTransform.Transform(new[] {points[i].X, points[i].Y});
```

```
            points[i] = new Geometries.Point(dblPoint[0], dblPoint[1]);
        }
    }
}
return points;
}
```

11.3.3.13　GetFootprint

得到图像四个角点组成的多边形对象，代码如下：

```
public Polygon GetFootprint()
{
    LinearRing myRing = new LinearRing(GetFourCorners());
    return new Polygon(myRing);
}
```

11.3.3.14　GetOneToOne

将图层缩放至原有的分辨率。首先调用_imagesize 的 Width、Height 属性，获取数据集的宽度和高度，分别保存于变量 DsWidth、DsHeight 中；调用 map.Envelope 属性获取地图的外包框矩形，赋予变量 bbox，调用 map.Size 获取地图的大小，赋予变量 size；然后比较 bbox 和_envelope 的边界值，计算图像要呈现的边界，保存于变量 left、top、right、bottom 中，代码如下：

```
public double GetOneToOne(Map map)
{
    double DsWidth = _imagesize.Width;
    double DsHeight = _imagesize.Height;
    double left, top, right, bottom;
    double dblImgEnvW, dblImgEnvH, dblWindowGndW, dblWindowGndH, dblImginMapW, dblImginMapH;
    BoundingBox bbox = map.Envelope;
    Size size = map.Size;
    // bounds of section of image to be displayed
    left = Math.Max(bbox.Left, _envelope.Left);
    top = Math.Min(bbox.Top, _envelope.Top);
    right = Math.Min(bbox.Right, _envelope.Right);
    bottom = Math.Max(bbox.Bottom, _envelope.Bottom);
```

计算经过地理变换后图像的宽度和高度（通过_envelope 成员变量计算），保存于变量 dblImgEnvW、dblImgEnvH；计算要呈现的窗口的宽度和高度（通过刚刚保存的 bbox 变量计算），保存于变量 dblWindowGndW、dblWindowGndH 中；计算在像素单位下，经过地理变换后的图像的高度和宽度，保存于变量 dblImginMapW、dblImginMapH 中；若图像是翻转的，返回 map.Zoom * (dblImginMapW / DsWidth)；若不是翻转的返回 map.Zoom * (dblImginMapH / DsWidth)：

11.3 GdalRasterLayer 类

```
        dblImgEnvW = _envelope.Right - _envelope.Left;
        dblImgEnvH = _envelope.Top - _envelope.Bottom;
        dblWindowGndW = bbox.Right - bbox.Left;
        dblWindowGndH = bbox.Top - bbox.Bottom;
        dblImginMapW = size.Width * (dblImgEnvW / dblWindowGndW);
        dblImginMapH = size.Height * (dblImgEnvH / dblWindowGndH);
        if (((dblImginMapW > dblImginMapH && DsWidth > DsHeight) ||
            (dblImginMapW < dblImginMapH && DsWidth < DsHeight))
            return map.Zoom * (dblImginMapW / DsWidth);
        else
            return map.Zoom * (dblImginMapH / DsWidth);
}
```

11.3.3.15 GetProjection

获取投影。这里先构建一个坐标系统工厂类 CoordinateSystemFactory 对象 cFac,用于创建坐标系统,代码如下:

```
        CoordinateSystemFactory cFac = new CoordinateSystemFactory();
```

若投影字符串变量的值不为空,就调用坐标工厂类的 CreateFromWkt 方法从字符串创建坐标系统并返回,否则返回 null,完整代码如下:

```
public ICoordinateSystem GetProjection()
{
        CoordinateSystemFactory cFac = new CoordinateSystemFactory();
        try
        {
            if (_projectionWkt! = "") return cFac.CreateFromWkt(_projectionWkt);
        }
        catch (Exception)
        {
        }
        return null;
}
```

11.3.3.16 GetTransform

获取变换矩阵,参数为需要变换到的投影对象。这里也是先创建一个坐标系统工厂类,并且利用工厂类创建源坐标系统对象,代码如下:

```
        CoordinateSystemFactory cFac = new CoordinateSystemFactory();
        ICoordinateSystem srcCoord = cFac.CreateFromWkt(_projectionWkt);
```

对源坐标系统和目的坐标系统检查完成后,调用坐标变换工厂对象的 CreateFromCoordinateSystems 方法,得到变换矩阵,保存到成员变量 _transform 中:

```
        _transform = new CoordinateTransformationFactory().CreateFromCoordinateSystems(srcCoord, tgt-
```

Coord);

完整代码如下:

```
private void GetTransform(ICoordinateSystem mapProjection)
{
    if (mapProjection = = null || _projectionWkt = = " ")
    {
        _transform = null;
        return;
    }
    CoordinateSystemFactory cFac = new CoordinateSystemFactory();
    // get our two projections
    ICoordinateSystem srcCoord = cFac.CreateFromWkt(_projectionWkt);
    ICoordinateSystem tgtCoord = mapProjection;
    // raster and map are in same projection, no need to transform
    if (srcCoord.WKT = = tgtCoord.WKT)
    {
        _transform = null;
        return;
    }
    // create transform
    _transform = new CoordinateTransformationFactory().CreateFromCoordinateSystems(srcCoord, tgtCoord);
}
```

11.3.3.17 GetZoomNearestRSet

获取缩放至 tiff 格式最相近的缩放水平的值。此方法前面部分的操作与 GetOneToOne 方法相同,不过这里不是返回缩放水平值,而是设置 dblTempWidth 值,代码如下:

```
public double GetZoomNearestRSet(Map map, bool bZoomIn)
{
    double DsWidth = _imagesize.Width;
    double DsHeight = _imagesize.Height;
    double left, top, right, bottom;
    double dblImgEnvW, dblImgEnvH, dblWindowGndW, dblWindowGndH, dblImginMapW, dblImginMapH;
    double dblTempWidth = 0;
    BoundingBox bbox = map.Envelope;
    Size size = map.Size;
    left = Math.Max(bbox.Left, _envelope.Left);
    top = Math.Min(bbox.Top, _envelope.Top);
    right = Math.Min(bbox.Right, _envelope.Right);
    bottom = Math.Max(bbox.Bottom, _envelope.Bottom);
```

```
dblImgEnvW = _envelope.Right - _envelope.Left;
dblImgEnvH = _envelope.Top - _envelope.Bottom;
dblWindowGndW = bbox.Right - bbox.Left;
dblWindowGndH = bbox.Top - bbox.Bottom;
dblImginMapW = size.Width * (dblImgEnvW / dblWindowGndW);
dblImginMapH = size.Height * (dblImgEnvH / dblWindowGndH);
if ((dblImginMapW > dblImginMapH && DsWidth > DsHeight) ||
    (dblImginMapW < dblImginMapH && DsWidth < DsHeight))
    dblTempWidth = dblImginMapW;
else
    dblTempWidth = dblImginMapH;
```

若缩放水平在显示的范围内，则根据 bZoomIn 的值，分别计算放大、缩小时的缩放比例值并返回：

```
if (DsWidth > dblTempWidth && (DsWidth / Math.Pow(2, 8)) < dblTempWidth)
{
    if (bZoomIn)
    {
        for (int i = 0; i <= 8; i++)
        {
            if (DsWidth / Math.Pow(2, i) > dblTempWidth)
            {
                if (DsWidth / Math.Pow(2, i + 1) < dblTempWidth)
                    return map.Zoom * (dblTempWidth / (DsWidth / Math.Pow(2, i)));
            }
        }
    }
    else
    {
        for (int i = 8; i >= 0; i--)
        {
            if (DsWidth / Math.Pow(2, i) < dblTempWidth)
            {
                if (DsWidth / Math.Pow(2, i - 1) > dblTempWidth)
                    return map.Zoom * (dblTempWidth / (DsWidth / Math.Pow(2, i)));
            }
        }
    }
}
return map.Zoom;
}
```

11.3.3.18　Render

渲染地图，代码如下：

```
public override void Render(Graphics g, Map map)
{
    if (disposed)
        throw (new ApplicationException("Error: An attempt was made to render a
            disposed layer"));
    GetPreview(_gdalDataset, map.Size, g, map.Envelope, null, map);
    base.Render(g, map);
}
```

11.3.3.19 ReprojectToMap

对地图使用投影变换，代码如下：

```
public void ReprojectToMap(Map map)
{
    GetTransform(null);
    ApplyTransformToEnvelope();
}
```

11.3.3.20 ResetHistoRectangle

重新获取直方图边界，代码如下：

```
public void ResetHistoRectangle()
{
    _histoBounds = new Rectangle((int)_envelope.Left, (int)_envelope.Bottom,
        (int)_envelope.Width, (int)_envelope.Height);
}
```

11.3.3.21 WritePixel

设置指定像素的值，参数为要设置的要素 X 坐标值 x、要设置的值 intVal、像素的大小 iPixelSize、色彩通道数组 ch 和要设置的像素所在行的指针 row。这里需要色彩的类型进行写入，分别为黑白图像、红外图像、彩红外图像、RGB 图像，代码如下：

```
protected unsafe void WritePixel(double x, double[] intVal, int iPixelSize, int[] ch, byte* row)
{
```

计算每个像素的偏移值，保存在整形变量 offsetX 中：

```
Int32 offsetX = (int)Math.Round(x) * iPixelSize;
```

当波段数为 1，位深不为 32，且 inval[0] 的值为 0 时，设置像素为红色：

```
if (Bands == 1 && _bitDepth != 32)
{
    if (ch[0] < 4)
    {
        if (_showClip)
        {
```

11.3 GdalRasterLayer 类

```
if (intVal[0] == 0)
{
    row[offsetX++] = 255;
    row[offsetX++] = 0;
    row[offsetX] = 0;
}
```

当波段数为1，位深不为32，且 inval [0] 的值为255时，设置像素为蓝色：

```
else if (intVal[0] == 255)
{
    row[offsetX++] = 0;
    row[offsetX++] = 0;
    row[offsetX] = 255;
}
```

其他情况设置为白色：

```
else
{
            row[offsetX++] = (byte)intVal[0];
            row[offsetX++] = (byte)intVal[0];
            row[offsetX] = (byte)intVal[0];
}
}
```

否则设置为 intVal 中存储的值：

```
else
{
    row[offsetX++] = (byte)intVal[0];
    row[offsetX++] = (byte)intVal[1];
    row[offsetX] = (byte)intVal[2];
}
}
```

若为红外灰度图，判断 intVal [3] 的值，若为0，设置为红色：

```
// IR grayscale
else if (DisplayIR && Bands == 4)
{
    for (int i = 0; i < Bands; i++)
    {
        if (ch[i] == 3)
        {
            if (_showClip)
            {
```

```
                    if ( intVal[3] = = 0 )
                    {
                        row[ (int)Math.Round(x) * iPixelSize ] = 255;
                        row[ (int)Math.Round(x) * iPixelSize + 1 ] = 0;
                        row[ (int)Math.Round(x) * iPixelSize + 2 ] = 0;
                    }
```

若为 255，设置为蓝色：

```
                    else if ( intVal[3] = = 255 )
                    {
                        row[ (int)Math.Round(x) * iPixelSize ] = 0;
                        row[ (int)Math.Round(x) * iPixelSize + 1 ] = 0;
                        row[ (int)Math.Round(x) * iPixelSize + 2 ] = 255;
                    }
```

否则设置为 intVal 中存储的值：

```
                    else
                    {
                        row[ (int)Math.Round(x) * iPixelSize ] = (byte)intVal[i];
                        row[ (int)Math.Round(x) * iPixelSize + 1 ] = (byte)intVal[i];
                        row[ (int)Math.Round(x) * iPixelSize + 2 ] = (byte)intVal[i];
                    }
                }
```

其他情况下，都设置为 intval 中存储的值：

```
                else
                {
                    row[ (int)Math.Round(x) * iPixelSize ] = (byte)intVal[i];
                    row[ (int)Math.Round(x) * iPixelSize + 1 ] = (byte)intVal[i];
                    row[ (int)Math.Round(x) * iPixelSize + 2 ] = (byte)intVal[i];
                }
            }
            Else continue;
        }
    }
    // CIR
```

若为彩红外波段：

```
    else if ( DisplayCIR && Bands = = 4 )
    {
        if ( _showClip )
        {
            if ( intVal[0] = = 0 && intVal[1] = = 0 && intVal[3] = = 0 )
```

```
            intVal[3] = intVal[0] = 0;
            intVal[1] = 255;
        }
        else if (intVal[0] == 255 && intVal[1] == 255 && intVal[3] == 255)
            intVal[1] = intVal[0] = 0;
    }
```

遍历每个波段，设置为 intVal 中存储的值：

```
    for (int i = 0; i < Bands; i++)
    {
        if (ch[i] != 0 && ch[i] != -1)
            row[(int)Math.Round(x) * iPixelSize + ch[i] - 1] = (byte)intVal[i];
    }
}
// RGB
else
{
    if (_showClip)
    {
        if (intVal[0] == 0 && intVal[1] == 0 && intVal[2] == 0)
        {
            intVal[0] = intVal[1] = 0;
            intVal[2] = 255;
        }
        else if (intVal[0] == 255 && intVal[1] == 255 && intVal[2] == 255)
            intVal[1] = intVal[2] = 0;
    }
    for (int i = 0; i < 3; i++)
    {
        if (ch[i] != 3 && ch[i] != -1)
            row[(int)Math.Round(x) * iPixelSize + ch[i]] = (byte)intVal[i];
    }
}
```

11.3.3.22 GetNonRotatedPreview

生成栅格数据的无旋转显示图。生成无旋转显示图的过程与生成显示图过程相似，只不过这里不使用图像旋转，速度会更快，代码如下：

```
private void GetNonRotatedPreview(Dataset dataset, Size size, Graphics g,
        BoundingBox bbox, ICoordinateSystem mapProjection)
{
    double[] geoTrans = new double[6];
```

```csharp
dataset.GetGeoTransform(geoTrans);
// default transform
if (!_useRotation && !_haveSpot || (geoTrans[0] == 0 && geoTrans[3] == 0))
    geoTrans = new[] { 999.5, 1, 0, 1000.5, 0, -1 };
Bitmap bitmap = null;
_geoTransform = new GeoTransform(geoTrans);
int DsWidth = 0;
int DsHeight = 0;
BitmapData bitmapData = null;
double[] intVal = new double[Bands];
int p_indx;
double bitScalar = 1.0;
double dblImginMapW = 0, dblImginMapH = 0, dblLocX = 0, dblLocY = 0;
int iPixelSize = 3; //Format24bppRgb = byte[b,g,r]
if (dataset != null)
{
    //check if image is in bounding box
    if ((bbox.Left > _envelope.Right) || (bbox.Right < _envelope.Left)
        || (bbox.Top < _envelope.Bottom) || (bbox.Bottom > _envelope.Top))
        return;
    DsWidth = _imagesize.Width;
    DsHeight = _imagesize.Height;
    _histogram = new List<int[]>();
    for (int i = 0; i < _lbands + 1; i++)
        _histogram.Add(new int[256]);
    double left = Math.Max(bbox.Left, _envelope.Left);
    double top = Math.Min(bbox.Top, _envelope.Top);
    double right = Math.Min(bbox.Right, _envelope.Right);
    double bottom = Math.Max(bbox.Bottom, _envelope.Bottom);
    double x1 = Math.Abs(_geoTransform.PixelX(left));
    double y1 = Math.Abs(_geoTransform.PixelY(top));
    double imgPixWidth = _geoTransform.PixelXwidth(right - left);
    double imgPixHeight = _geoTransform.PixelYwidth(bottom - top);
    //get screen pixels image should fill
    double dblBBoxW = bbox.Right - bbox.Left;
    double dblBBoxtoImgPixX = imgPixWidth / dblBBoxW;
    dblImginMapW = size.Width * dblBBoxtoImgPixX *
        _geoTransform.HorizontalPixelResolution;
    double dblBBoxH = bbox.Top - bbox.Bottom;
    double dblBBoxtoImgPixY = imgPixHeight / dblBBoxH;
    dblImginMapH = size.Height * dblBBoxtoImgPixY *
        -_geoTransform.VerticalPixelResolution;
    if ((dblImginMapH == 0) || (dblImginMapW == 0))
```

11.3 GdalRasterLayer 类

```
        return;
// ratios of bounding box to image ground space
double dblBBoxtoImgX = size.Width / dblBBoxW;
double dblBBoxtoImgY = size.Height / dblBBoxH;
// set where to display bitmap in Map
if (bbox.Left != left)
{
    if (bbox.Right != right)
        dblLocX = (_envelope.Left - bbox.Left) * dblBBoxtoImgX;
    else
        dblLocX = size.Width - dblImginMapW;
}
if (bbox.Top != top)
{
    if (bbox.Bottom != bottom)
        dblLocY = (bbox.Top - _envelope.Top) * dblBBoxtoImgY;
    else
        dblLocY = size.Height - dblImginMapH;
}
// scale
if (_bitDepth == 12)
    bitScalar = 16.0;
else if (_bitDepth == 16)
    bitScalar = 256.0;
else if (_bitDepth == 32)
    bitScalar = 16777216.0;
try
{
    bitmap = new Bitmap((int)Math.Round(dblImginMapW),
        (int)Math.Round(dblImginMapH),
                    PixelFormat.Format24bppRgb);
    bitmapData =
            bitmap.LockBits(
                new Rectangle(0, 0, (int)Math.Round(dblImginMapW),
                (int)Math.Round(dblImginMapH)),
                ImageLockMode.ReadWrite, bitmap.PixelFormat);
    byte cr = _noDataInitColor.R;
    byte cg = _noDataInitColor.G;
    byte cb = _noDataInitColor.B;
    //
    Double[] noDataValues = new Double[Bands];
    Double[] scales = new Double[Bands];
    ColorTable colorTable = null;
```

```csharp
unsafe
{
    double[][] buffer = new double[Bands][];
    Band[] band = new Band[Bands];
    int[] ch = new int[Bands];
    // get data from image
    for (int i = 0; i < Bands; i++)
    {
        buffer[i] = new double[(int)Math.Round(dblImginMapW) *
            (int)Math.Round(dblImginMapH)];
        band[i] = dataset.GetRasterBand(i + 1);
        //get nodata value if present
        Int32 hasVal = 0;
        band[i].GetNoDataValue(out noDataValues[i], out hasVal);
        if (hasVal == 0) noDataValues[i] = Double.NaN;
        band[i].GetScale(out scales[i], out hasVal);
        if (hasVal == 0) scales[i] = 1.0;
        band[i].ReadRaster((int)Math.Round(x1),
            (int)Math.Round(y1), (int)Math.Round(imgPixWidth),
            (int)Math.Round(imgPixHeight), buffer[i],
            (int)Math.Round(dblImginMapW),
            (int)Math.Round(dblImginMapH), 0, 0);
        if (band[i].GetRasterColorInterpretation() ==
            ColorInterp.GCI_BlueBand) ch[i] = 0;
        else if (band[i].GetRasterColorInterpretation() ==
            ColorInterp.GCI_GreenBand) ch[i] = 1;
        else if (band[i].GetRasterColorInterpretation() ==
            ColorInterp.GCI_RedBand) ch[i] = 2;
        else if (band[i].GetRasterColorInterpretation() ==
            ColorInterp.GCI_Undefined)
        {
            if (Bands > 1)
                ch[i] = 3; // infrared
            else
            {
                ch[i] = 4;
                if (_colorBlend == null)
                {
                    Double dblMin, dblMax;
                    band[i].GetMinimum(out dblMin, out hasVal);
                    if (hasVal == 0) dblMin = Double.NaN;
                    band[i].GetMaximum(out dblMax, out hasVal);
                    if (hasVal == 0) dblMax = double.NaN;
```

11.3 GdalRasterLayer 类

```
                    if (Double.IsNaN(dblMin) ||
                    Double.IsNaN(dblMax))
                    {
                        double dblMean, dblStdDev;
                        band[i].GetStatistics(0, 1, out dblMin, out
                        dblMax, out dblMean, out dblStdDev);
                        //double dblRange = dblMax - dblMin;
                        //dblMin -= 0.1 * dblRange;
                        //dblMax += 0.1 * dblRange;
                    }
                    Single[] minmax = new float[]
                    {Convert.ToSingle(dblMin), 0.5f *
                     Convert.ToSingle(dblMin + dblMax),
                     Convert.ToSingle(dblMax) };
                    Color[] colors = new Color[] { Color.Blue,
                    Color.Yellow, Color.Red };
                    _colorBlend = new ColorBlend(colors, minmax);
                }
                intVal = new Double[3];
            }
        }
        else if (band[i].GetRasterColorInterpretation() ==
        ColorInterp.GCI_GrayIndex) ch[i] = 0;
        else if (band[i].GetRasterColorInterpretation() ==
        ColorInterp.GCI_PaletteIndex)
        {
            colorTable = band[i].GetRasterColorTable();
            ch[i] = 5;
            intVal = new Double[3];
        }
        else ch[i] = -1;
    }
    if (_bitDepth == 32)
        ch = new[] { 0, 1, 2 };
    p_indx = 0;
    for (int y = 0; y < Math.Round(dblImginMapH); y++)
    {
        byte* row = (byte*)bitmapData.Scan0 + (y * bitmapData.Stride);
        for (int x = 0; x < Math.Round(dblImginMapW); x++, p_indx++)
        {
            for (int i = 0; i < Bands; i++)
            {
                intVal[i] = buffer[i][p_indx]/bitScalar;
```

```
                        Double imageVal = intVal[i] = intVal[i]/bitScalar;
                        if (ch[i] == 4)
                        {
                            if (imageVal != noDataValues[i])
                            {
                                Color color =
                        _colorBlend.GetColor(Convert.ToSingle(imageVal));
                                intVal[0] = color.B;
                                intVal[1] = color.G;
                                intVal[2] = color.R;
                                //intVal[3] = ce.c4;
                            }
                            else
                            {
                                intVal[0] = cb;
                                intVal[1] = cg;
                                intVal[2] = cr;
                            }
                        }
                        else if (ch[i] == 5 && colorTable != null)
                        {
                            if (imageVal != noDataValues[i])
                            {
                                ColorEntry ce =
                        colorTable.GetColorEntry(Convert.ToInt32(imageVal));
                                intVal[0] = ce.c3;
                                intVal[1] = ce.c2;
                                intVal[2] = ce.c1;
                                //intVal[3] = ce.c4;
                            }
                            else
                            {
                                intVal[0] = cb;
                                intVal[1] = cg;
                                intVal[2] = cr;
                            }
                        }
                        else
                        {
                            if (_colorCorrect)
                            {
                                intVal[i] = ApplyColorCorrection(intVal[i],
                        0, ch[i], 0, 0);
```

```
                    if ( _lbands > = 3)
                        _histogram[_lbands][
                        (int) (intVal[2] * 0.2126 +
                        intVal[1] * 0.7152 + intVal[0] * 0.0722)] + + ;
                    }
                }
                if (intVal[i] > 255)
                    intVal[i] = 255;
            }
            WritePixel(x, intVal, iPixelSize, ch, row);
        }
    }
}
catch
{
    return;
}
finally
{
    if (bitmapData ! = null)
        bitmap.UnlockBits(bitmapData);
}
}
if (_transparentColor ! = Color.Empty)
    bitmap.MakeTransparent(_transparentColor);
g.DrawImage(bitmap, new Point((int)Math.Round(dblLocX),
    (int)Math.Round(dblLocY)));
}
```

11.3.3.23 GetPreview

生成栅格数据的显示图,代码如下:

```
protected virtual void GetPreview(Dataset dataset, Size size, Graphics g,
BoundingBox displayBbox, ICoordinateSystem mapProjection, Map map)
{
```

构建第一个 double 类型的数组,用于存放仿射变换矩阵,将其作为参数调用数据集的 GetGeoTransform 方法,获取其值:

```
double[] geoTrans = new double[6];
_gdalDataset.GetGeoTransform(geoTrans);
```

根据是否使用旋转,生成不同的图像,若不使用旋转,就调用 GetNonRotatedPreview 方法生成无旋转图像,然后直接返回:

```
// not rotated, use faster display method
if ((! _useRotation ||
    (geoTrans[1] == 1 && geoTrans[2] == 0 && geoTrans[4] == 0 && Math.Abs(geoTrans[5]) == 1))
            && ! _haveSpot && _transform == null)
{
    GetNonRotatedPreview(dataset, size, g, displayBbox, mapProjection);
    return;
}
```

若需要旋转并且有污点,就是用默认的旋转数组:

```
// not rotated, but has spot... need default rotation
else if ((geoTrans[0] == 0 && geoTrans[3] == 0) && _haveSpot)
    geoTrans = new[] { 999.5, 1, 0, 1000.5, 0, -1 };
```

生成集合变换对象,保存于成员变量_geoTransform 中:

```
_geoTransform = new GeoTransform(geoTrans);
```

构建变量 DsWidth、DsHeight,分别用于保存图像的宽度、长度:

```
double DsWidth = _imagesize.Width;
double DsHeight = _imagesize.Height;
```

变量 left、top、right、bottom 分别表示图像的左、上、右、下边界值:

```
double left, top, right, bottom;
```

变量 GndX、GndY、ImgX、ImgY、PixX、PixY 分别表示点的投影坐标 X 值、点的投影坐标 Y 值、点的图像坐标 X 值、点的图像坐标 Y 值、要显示的图片的坐标 X 值、要显示的图片的坐标 Y 值:

```
double GndX = 0, GndY = 0, ImgX = 0, ImgY = 0, PixX, PixY;
```

数组变量 intVal 用于存储图像具体的值(强制转换为整形):

```
double[] intVal = new double[Bands];
```

变量 imageVal、SpotVal、bitScalar 分别表示图像值、污点值、位幅:

```
double imageVal = 0, SpotVal = 0;
double bitScalar = 1.0;
```

变量 bitmap 表示要绘制的图像:

```
Bitmap bitmap = null;
```

变量 bitmapTL、bitmapBR 分别表示要绘制的图像的左上角点、右下角点:

```
Point bitmapTL = new Point(), bitmapBR = new Point();
Geometries.Point imageTL = new Geometries.Point(), imageBR = new Geometries.Point();
```

11.3 GdalRasterLayer 类

变量 shownImageBbox、trueImageBbox 分别表示图像的显示外包框矩形及实际外框矩形：

```
BoundingBox shownImageBbox, trueImageBbox;
```

变量 bitmapLength、bitmapHeight、displayImageLength、displayImageHeight 分别表示要绘制的图像的长度、高度，要显示的图像的长度、高度：

```
int bitmapLength, bitmapHeight;
int displayImageLength, displayImageHeight;
```

整形变量 iPixelSize 表示像素的大小为 3：

```
int iPixelSize = 3; //Format24bppRgb = byte[b,g,r]
```

当数据集不为空时：

```
if (dataset ! = null)
{
```

检查要显示的图像是否超出了真实图像的范围，若超出直接返回：

```
//check if image is in bounding box
if (((displayBbox.Left > _envelope.Right) || (displayBbox.Right < _envelope.Left)
    | (displayBbox.Top < _envelope.Bottom) || (displayBbox.Bottom > _envelope.Top))
        return;
```

初始化统计直方图数组列表，每个列表中包含一个长度为 256 的数组：

```
// init histo
_histogram = new List<int[]>();
for (int i = 0; i < _lbands + 1; i++)
    _histogram.Add(new int[256]);
```

计算要显示的图像块的边界值，保存于变量 left、top、right、bottom 中：

```
// bounds of section of image to be displayed
left = Math.Max(displayBbox.Left, _envelope.Left);
top = Math.Min(displayBbox.Top, _envelope.Top);
right = Math.Min(displayBbox.Right, _envelope.Right);
bottom = Math.Max(displayBbox.Bottom, _envelope.Bottom);
```

构建真实的图像块的外包框：

```
trueImageBbox = new BoundingBox(left, bottom, right, top);
```

把真实图像块的外包框的投影转换到当前的投影：

```
// put display bounds into current projection
if (_transform ! = null)
{
    _transform.MathTransform.Invert();
```

```
            shownImageBbox = GeometryTransform.TransformBox(trueImageBbox, _transform.MathTransform);
                    _transform.MathTransform.Invert();
        }
        else shownImageBbox = trueImageBbox;
```

计算要显示图像边界点, 保存在变量 imageBR 和 ImageTL 中:

```
        // find min/max x and y pixels needed from image
        imageBR.X = (int)(Math.Max(_geoTransform.GroundToImage(shownImageBbox.TopLeft).X,
                Math.Max(_geoTransform.GroundToImage(shownImageBbox.TopRight).X,
                Math.Max(_geoTransform.GroundToImage(shownImageBbox.BottomLeft).X,
                _geoTransform.GroundToImage(shownImageBbox.BottomRight).X))) + 1);
        imageBR.Y = (int)(Math.Max(_geoTransform.GroundToImage(shownImageBbox.TopLeft).Y,
                Math.Max(_geoTransform.GroundToImage(shownImageBbox.TopRight).Y,
                Math.Max(_geoTransform.GroundToImage(shownImageBbox.BottomLeft).Y,
                _geoTransform.GroundToImage(shownImageBbox.BottomRight).Y))) + 1);
        imageTL.X = (int)Math.Min(_geoTransform.GroundToImage(shownImageBbox.TopLeft).X,
                Math.Min(_geoTransform.GroundToImage(shownImageBbox.TopRight).X,
                Math.Min(_geoTransform.GroundToImage(shownImageBbox.BottomLeft).X,
                _geoTransform.GroundToImage(shownImageBbox.BottomRight).X)));
        imageTL.Y = (int)Math.Min(_geoTransform.GroundToImage(shownImageBbox.TopLeft).Y,
                Math.Min(_geoTransform.GroundToImage(shownImageBbox.TopRight).Y,
                Math.Min(_geoTransform.GroundToImage(shownImageBbox.BottomLeft).Y,
                _geoTransform.GroundToImage(shownImageBbox.BottomRight).Y)));
```

要显示的图像的边界要在图像的大小范围内, 否则只显示图像范围内的:

```
        // stay within image
        if (imageBR.X > _imagesize.Width) imageBR.X = _imagesize.Width;
        if (imageBR.Y > _imagesize.Height) imageBR.Y = _imagesize.Height;
        if (imageTL.Y < 0) imageTL.Y = 0;
        if (imageTL.X < 0) imageTL.X = 0;
```

计算要显示的图像的宽度、高度:

```
        displayImageLength = (int)(imageBR.X - imageTL.X);
        displayImageHeight = (int)(imageBR.Y - imageTL.Y);
```

调用投影变换对象的 ImageToGround 方法, 将要显示的图像的边界点坐标转换为对应的投影坐标:

```
        // find ground coordinates of image pixels
        Geometries.Point groundBR = _geoTransform.ImageToGround(imageBR);
        Geometries.Point groundTL = _geoTransform.ImageToGround(imageTL);
```

将真实的图像外包框角点转换为图片坐标, 保存于图像右下角点 bitmapBR、左上角点 bitmapTL 中:

```
// convert ground coordinates to map coordinates to figure out where to place the bitmap
bitmapBR = new Point((int)map.WorldToImage(trueImageBbox.BottomRight).X + 1,
                    (int)map.WorldToImage(trueImageBbox.BottomRight).Y + 1);
bitmapTL = new Point((int)map.WorldToImage(trueImageBbox.TopLeft).X,
                    (int)map.WorldToImage(trueImageBbox.TopLeft).Y);
```

计算图像的宽度、高度:

```
bitmapLength = bitmapBR.X - bitmapTL.X;
bitmapHeight = bitmapBR.Y - bitmapTL.Y;
```

检查要显示的图像的大小是否在图像的大小范围内:

```
// check to see if image is on its side
if(bitmapLength > bitmapHeight && displayImageLength < displayImageHeight)
{
    displayImageLength = bitmapHeight;
    displayImageHeight = bitmapLength;
}
else
{
    displayImageLength = bitmapLength;
    displayImageHeight = bitmapHeight;
}
```

根据不同的位深,赋予不同的位宽值(用于计算对应的十进制值):

```
// scale
if(_bitDepth == 12) bitScalar = 16.0;
else if(_bitDepth == 16) bitScalar = 256.0;
else if(_bitDepth == 32) bitScalar = 16777216.0;
// 0 pixels in length or height, nothing to display
if(bitmapLength < 1 || bitmapHeight < 1) return;
```

实例化一个栅格图像对象 bitmap、图像数据对象 bitmapData、用于绘制栅格数据:

```
//initialize bitmap
bitmap = new Bitmap(bitmapLength, bitmapHeight, PixelFormat.Format24bppRgb);
    BitmapData bitmapData = bitmap.LockBits(new Rectangle(0, 0, bitmapLength, bitmapHeight), ImageLockMode.ReadWrite, bitmap.PixelFormat);
try
{
    unsafe
    {
```

获取无数据点的默认 R、G、B 值:

```
// turn everything to _noDataInitColor, so we can make fill transparent
```

```
byte cr = _noDataInitColor.R;
byte cg = _noDataInitColor.G;
byte cb = _noDataInitColor.B;
```

设置 bitmap 图像的所有像素的值为默认值,按 B、G、R 顺序赋值:

```
for(int y = 0; y < bitmapHeight; y++)
{
    byte * brow = (byte * )bitmapData.Scan0 + (y * bitmapData.Stride);
    for(int x = 0; x < bitmapLength; x++)
    {
        Int32 offsetX = x * 3;
        brow[offsetX++] = cb;B
        brow[offsetX++] = cg;G
        brow[offsetX] = cr;R
    }
}
```

构建一个二维数组 tempBuffe 用于缓存波段数据,三位数组 buffer 用于缓存每个波段的数据:

```
// create 3 dimensional buffer [band][x pixel][y pixel]
double[][] tempBuffer = new double[Bands][];
double[][][] buffer = new double[Bands][][];
```

设置 buffer 中的第二维大小为要显示的图像的高度:

```
for(int i = 0; i < Bands; i++)
{
    buffer[i] = new double[displayImageLength][];
    for(int j = 0; j < displayImageLength; j++)
        buffer[i][j] = new double[displayImageHeight];
}
```

新建一个波段数组 band,用于保存波段;整形数组 ch 用于保存波段值:

```
Band[] band = new Band[Bands];
int[] ch = new int[Bands];
```

Double 类型数组 noDataValues、scales 分别保存每个波段的无数据值及每个波段的幅度范围:

```
//
Double[] noDataValues = new Double[Bands];
Double[] scales = new Double[Bands];
ColorTable colorTable = null;
```

遍历每个波段,获取数据:

11.3 GdalRasterLayer 类

```
// get data from image
for ( int i = 0; i < Bands; i ++ )
{
```

tempBuffer 的第二维的大小设置为图像的高度乘以宽度：

```
tempBuffer[i] = new double[displayImageLength * displayImageHeight];
```

调用数据集的 GetRasterBand 方法获取波段：

```
band[i] = dataset.GetRasterBand(i + 1);
```

获取波段中 NoData 的值，保存于变量 hasVal 中，若没则设置为 Double.NaN：

```
//get nodata value if present
Int32 hasVal = 0;
band[i].GetNoDataValue( out noDataValues[i], out hasVal);
if ( hasVal == 0 ) noDataValues[i] = Double.NaN;
```

获取每个波段的范围，保存在 scale 数组中，默认设置为 1.0：

```
band[i].GetScale( out scales[i], out hasVal);
if ( hasVal == 0 ) scales[i] = 1.0;
```

调用波段的 ReadRaster 方法读取波段数据，保存在 tempBuffer 数组中：

```
band[i].ReadRaster( (int)imageTL.X, (int)imageTL.Y, (int)(imageBR.X - imageTL.X),
(int)
(imageBR.Y - imageTL.Y), tempBuffer[i], displayImageLength, displayImageHeight, 0, 0);
```

将 tempBuffer 数组中的值存储到 buffer 三维缓冲数组中：

```
// parse temp buffer into the image x y value buffer
long pos = 0;
for ( int y = 0; y < displayImageHeight; y ++ )
{
    for ( int x = 0; x < displayImageLength; x ++, pos ++ )
        buffer[i][x][y] = tempBuffer[i][pos];
}
```

调用波段的 GetRasterColorInterpretation 方法获取色彩通道的组合，蓝色波段设置值为 0，绿色波段设置值为 1，红色波段设置值为 2：

```
if (band[i].GetRasterColorInterpretation() == ColorInterp.GCI_BlueBand) ch[i] = 0;
else if (band[i].GetRasterColorInterpretation() == ColorInterp.GCI_GreenBand) ch[i] = 1;
else if (band[i].GetRasterColorInterpretation() == ColorInterp.GCI_RedBand) ch[i] = 2;
else if (band[i].GetRasterColorInterpretation() == ColorInterp.GCI_Undefined)
{
```

未定义情况下，若波段数大于1，则设置值为3，表示近红，否则设置值为4：

```
    if (Bands > 1) ch[i] = 3; // infrared
    else
    {
        ch[i] = 4;
```

当色彩混合变量的值为空时，进行色彩混合，这里先获取波段的最大最小值，然后构造一个 ColorBlend（RGB 组合）对象，赋予_colorBlend：

```
    if (_colorBlend == null)
    {
        Double dblMin, dblMax;
        band[i].GetMinimum(out dblMin, out hasVal);
        if (hasVal == 0) dblMin = Double.NaN;
        band[i].GetMaximum(out dblMax, out hasVal);
        if (hasVal == 0) dblMax = double.NaN;
        if (Double.IsNaN(dblMin) || Double.IsNaN(dblMax))
        {
            double dblMean, dblStdDev;
            band[i].GetStatistics(0, 1, out dblMin, out dblMax, out dblMean, out dblStdDev);
        }
        Single[] minmax = new float[]
    { Convert.ToSingle(dblMin), 0.5f * Convert.ToSingle(dblMin + dblMax), Convert.ToSingle(dblMax) };
        Color[] colors = new Color[] { Color.Blue, Color.Yellow, Color.Red };
        _colorBlend = new ColorBlend(colors, minmax);
    }
    intVal = new Double[3];
}
}
```

当为灰度索引时，设置值为 0：

```
    else if (band[i].GetRasterColorInterpretation() == ColorInterp.GCI_GrayIndex) ch[i] = 0;
```

当值为调色板索引时，调用波段的 GetRasterColorTable 方法获取色彩表，并设置值为 5：

```
    else if (band[i].GetRasterColorInterpretation() == ColorInterp.GCI_PaletteIndex)
    {
        colorTable = band[i].GetRasterColorTable();
        ch[i] = 5;
        intVal = new Double[3];
    }
```

剩余的情况设置值为 -1：

```
    else ch[i] = -1;
```

11.3 GdalRasterLayer 类

保存现有的变量的值:

```
// store these values to keep from having to make slow method calls
int bitmapTLX = bitmapTL.X;
int bitmapTLY = bitmapTL.Y;
double imageTop = imageTL.Y;
double imageLeft = imageTL.X;
double dblMapPixelWidth = map.PixelWidth;
double dblMapPixelHeight = map.PixelHeight;
double dblMapMinX = map.Envelope.Min.X;
double dblMapMaxY = map.Envelope.Max.Y;
double geoTop, geoLeft, geoHorzPixRes, geoVertPixRes, geoXRot, geoYRot;
// get inverse values
geoTop = _geoTransform.Inverse[3];
geoLeft = _geoTransform.Inverse[0];
geoHorzPixRes = _geoTransform.Inverse[1];
geoVertPixRes = _geoTransform.Inverse[5];
geoXRot = _geoTransform.Inverse[2];
geoYRot = _geoTransform.Inverse[4];
double dblXScale = (imageBR.X - imageTL.X) / (displayImageLength - 1);
double dblYScale = (imageBR.Y - imageTL.Y) / (displayImageHeight - 1);
double[] dblPoint;
// get inverse transform
// NOTE: calling transform.MathTransform.Inverse() once and storing it
// is much faster than having to call every time it is needed
IMathTransform inverseTransform = null;
if (_transform != = null)
    inverseTransform = _transform.MathTransform.Inverse();
```

遍历图像的每个像素,往 bitmapData 写入值:

```
for (PixY = 0; PixY < bitmapBR.Y - bitmapTL.Y; PixY++)
{
```

获取行指针:

```
byte * row = (byte *)bitmapData.Scan0 + ((int)Math.Round(PixY) * bitmapData.Stride);
for (PixX = 0; PixX < bitmapBR.X - bitmapTL.X; PixX++)
{
```

转换每个点的图像坐标到对应的投影坐标:

```
GndX = dblMapMinX + (PixX + bitmapTLX) * dblMapPixelWidth;
GndY = dblMapMaxY - (PixY + bitmapTLY) * dblMapPixelHeight;
```

当 _transform 不为空时,进行坐标变换:

```
// transform ground point if needed
if (_transform ! = null)
{
    dblPoint = inverseTransform.Transform(new[]{GndX, GndY});
    GndX = dblPoint[0];
    GndY = dblPoint[1];
}
```

将地理坐标转换为图像坐标,检查每个点是否超出图像的范围,若超出则跳出,继续下一次循环:

```
// same as GeoTransform.GroundToImage(), but much faster using stored values...
ImgX = (geoLeft + geoHorzPixRes * GndX + geoXRot * GndY);
ImgY = (geoTop + geoYRot * GndX + geoVertPixRes * GndY);

if (ImgX < imageTL.X || ImgX > imageBR.X || ImgY < imageTL.Y || ImgY > imageBR.Y)
    continue;
```

获取像素点对应的每个波段的值,并强制转化为整形值,存储于变量 intVal 数值中:

```
// color correction
for (int i = 0; i < Bands; i++)
{
    intVal[i] =
        buffer[i][(int)((ImgX - imageLeft) / dblXScale)][
            (int)((ImgY - imageTop) / dblYScale)];
```

将值转换为对应的十进制值:

```
imageVal = SpotVal = intVal[i] = intVal[i] / bitScalar;
```

当 ch 当前索引的值为 4 时,表示波段数小于或等于 1,当前单元的值不等于 NoData 值时,调用_colorBlend 获取色彩混合值,设置 intVal 数组的值分别为色彩的 B、R、G 值,否则设置为 NoData 的 B、G、R 值:

```
if (ch[i] == 4)
{
    if (imageVal != noDataValues[i])
    {
        Color color = _colorBlend.GetColor(Convert.ToSingle(imageVal));
        intVal[0] = color.B;
        intVal[1] = color.G;
        intVal[2] = color.R;
    }
    else
    {
        intVal[0] = cb;
```

11.3 GdalRasterLayer 类

```
        intVal[1] = cg;
        intVal[2] = cr;
    }
}
```

当 ch 的当前索引值为 5，且颜色表不为空时，调用色彩表的 GetColorEntry 获取调色板的值，由于调色板是 RGB 组合，c1、c2、c3 分别表示红（Red）、绿（Green）、蓝（Blue），设置 intVal 数组为对应值：

```
else if (ch[i] = = 5 && colorTable ! = null)
{
    if (imageVal ! = noDataValues[i])
    {
        ColorEntry ce = colorTable.GetColorEntry(Convert.ToInt32(imageVal));
        intVal[0] = ce.c3;
        intVal[1] = ce.c2;
        intVal[2] = ce.c1;
    }
    else
    {
        intVal[0] = cb;
        intVal[1] = cg;
        intVal[2] = cr;
    }
}
```

其他情况下：

```
else
{
```

若使用色彩纠正：

```
if (_colorCorrect)
{
```

设置 intVal 的值为调用 ApplyColorCorrection 方法进行色彩纠正后的值：

```
intVal[i] = ApplyColorCorrection(imageVal, SpotVal, ch[i], GndX, GndY);
    // if pixel is within ground boundary, add its value to the histogram
```

若此像素点在边界范围内，将当前点的统计直方图的数值自动加 1：

```
if (ch[i] ! = -1 && intVal[i] > 0 && (_histoBounds.Bottom > = (int)GndY &&
    _histoBounds.Top < = (int)GndY && _histoBounds.Left < = (int)GndX && _histo-
Bounds.Right > = (int)GndX)
{
    _histogram[ch[i]][(int)intVal[i]] + +;
```

若 intVal 的值大于 255，设置为 255：

```
if (intVal[i] > 255) intVal[i] = 255;
    }
    }
```

当波段数大于 3 时，做色彩变幻，调整亮度：

```
// luminosity
if (_lbands >= 3)
_histogram[_lbands][(int)(intVal[2] * 0.2126 + intVal[1] * 0.7152 + intVal[0] * 0.0722)]
++;
```

最后调用 WritePixel 方法写入值，参数为像素 X 坐标、值、像素大小、色彩、行指针：

```
WritePixel(PixX, intVal, iPixelSize, ch, row);
            }
        }
    }
}
finally
{
    bitmap.UnlockBits(bitmapData);
}
}
```

调用 bitmap 的 MakeTransparent 方法将 _noDataInitColor 设置为透明色：

```
bitmap.MakeTransparent(_noDataInitColor);
if (_transparentColor != Color.Empty)
    bitmap.MakeTransparent(_transparentColor);
```

最后调用 Graphic 的 DrawImage 绘制出图像：

```
g.DrawImage(bitmap, new Point(bitmapTL.X, bitmapTL.Y));
}
```

复习思考题

11-1 如何对 DataTablePoint 数据源进行扩展？以 DataTablePoint 为例思考对其他数据源的扩展。

11-2 理解 SharpMap 中扩展图层对象的方式。

附录　书中多次引用的基本概念

A　SRID（Spatial Reference System Identifier）

维基百科（http：//en. wikipedia. org）中对 SRID 有如下定义：A Spatial Reference System Identifier (SRID) is a unique value used to unambiguously identify projected, unprojected, and local spatial coordinate system definitions. These coordinate systems form the heart of all GIS applications。其基本意思是，SRID 是一套唯一编码，用以标识投影、非投影、局部坐标系，这些坐标系构成了 GIS 应用的核心。4326 常常被认为是地图投影中最简单的坐标系统，表达方式如下：

```
GEOGCS["WGS 84",
DATUM["WGS_1984",
SPHEROID["WGS 84",6378137,298.257223563,
AUTHORITY["EPSG","7030"]],
AUTHORITY["EPSG","6326"]],
PRIMEM["Greenwich",0,
AUTHORITY["EPSG","8901"]],
UNIT["degree",0.01745329251994328,
AUTHORITY["EPSG","9122"]],
AUTHORITY["EPSG","4326"]]
```

SharpMap 目录下有一个 SRID.csv 文件，列出了常用的空间参考系统。

B　GeoServer

GeoServer 是一个功能齐全、遵循 OGC 开放标准的开源 WFS-T 和 WMS 服务器。利用 GeoServer 可以把数据作为 maps/images 来发布（利用 WMS 来实现），也可以直接发布实际的数据（利用 WFS 来实现），同时也提供了修改、删除和新增的功能（利用 WFS-T）。

C　S-57

S-57 是数字化海道测量数据传输标准，旨在规范各国海道测量部门用于传输的数字海道数据。从模型、数据结构、S-57 文件（应用简档）、技术特点以及注意事项等方面阐述了 S-57 的精髓，为电子海图的生成、传输、交换以及 ECDIS 的应用提供了标准化规范。

D　PostgreSQL

PostgreSQL 是一种对象 - 关系型数据库管理系统（ORDBMS），也是目前功能最强大、特性最丰富和最复杂的自由软件数据库系统。它起源于伯克利（BSD）的数据库研究计划，目前是最重要的开源数据库产品开发项目之一，有着非常广泛的用户。PostGIS 在对象关系型数据库 PostgreSQL 上增加了存储管理空间数据的能力，相当于 Oracle 的 spatial 部分。PostGIS 最大的特点是符合并且实现了 OpenGIS 的一些规范，是最著名的开源 GIS 数据库。

E SQLite

SQLite 号称全世界最小的数据库，在几乎绝大多数数据库都具有空间数据的存储和查询功能后，SQLite 目前也有了空间数据支持的扩展，利用这个扩展，可以按照 OGC 的 Simple Feature Access 标准存取空间数据。这个项目名叫 SpatiaLite，与其一同开发的还有一个 VirtualShape。前者为 SQLite 增加空间数据支持，后者可以把一个 Shapefile 作为 SQLite 的数据库。

F Proj4. NET

Proj4. NET 是一套基于 C#开发的 .NET 类库，用于基准面、地理坐标系之间的转换，网址为 http://proj4net.codeplex.com/。

参考文献

[1] Shahab Fazal. GIS Basics [M]. New Delhi: New Age International Publishers, 2008.
[2] Michael DeMers. GIS for Dummies [M]. New York: Wiley Publishing, 2009.
[3] Markus Neteler. Open Source GIS – A Grass GIS Approach [M]. Hamburg: Springer, 2008.
[4] Scott Davis. GIS for Web Developers [M]. The Pragmatic Programmers LLC, 2007.
[5] Nick Randolph. Professional Visual Studio 2010 [M]. New York: Wiley Publishing, 2010.
[6] Christian Nage. Professional C# 4 and .NET 4 [M]. New York: Wiley Publishing, 2010.
[7] 维基百科. http://en.wikipedia.org/wiki/Main_Page.

冶金工业出版社部分图书推荐

书　名	作　者	定价(元)
C++程序设计（本科教材）	高　潮	40.00
C语言程序设计与实训（高职高专）	闻红军	30.00
Visual C++环境下Mapx的开发技术	尹旭日	26.00
计算机实用软件大全	何培民	159.00
轧制过程的计算机控制系统	赵　刚	25.00
计算机辅助建筑设计——建筑效果图设计教程	刘声远	25.00
AutoCAD项目式教程	陈胜利	28.50
Pro/E Wildfire中文版模具设计教程	张武军	39.00
粒子群优化算法	李　丽	20.00
Solid Works 2006零件与装配设计教程	岳荣刚	29.00
最优化原理与方法（修订版）（本科教材）	薛嘉庆	18.00
可编程序控制器及常用控制电器（第2版）（本科教材）	何友华	30.00
可编程序控制器原理及应用系统设计技术（第2版）（本科教材）	宋德玉	26.00
监控组态软件的设计与开发	李建伟	33.00
冶金熔体结构和性质的计算机模拟计算	谢　刚	20.00
材料成形计算机模拟（本科教材）	辛启斌	17.00
计算机病毒防治与信息安全知识300问	张　洁	25.00
土木工程材料（英文）（本科教材）	陈　瑜	27.00
FIDIC条件与合同管理（本科教材）	李明顺	38.00
建筑施工实训指南（高专教材）	韩玉文	28.00
建筑结构振动计算与抗振措施	张荣山	55.00
岩巷工程施工——掘进工程	孙延宗	120.00
岩巷工程施工——支护工程	孙延宗	100.00
钢骨混凝土异形柱	李　哲	25.00
地下工程智能反馈分析方法与应用	姜谙男	36.00
建筑环境工程设备基础	李绍勇	27.00
实用有色金属科技日语教程	王春香	33.00